Lecture Notes in Biomathematics

Managing Editor: S. Levin

62

Bruce J. West

An Essay on the Importance of Being Nonlinear

Springer-Verlag
Berlin Heidelberg New York Tokyo

Author

Bruce J. West
Center for Studies of Nonlinear Dynamics*, La Jolla Institute
3252 Holiday Court, Suite 208, La Jolla, California 92037, USA

* (affiliated with the University of California, San Diego)

Mathematics Subject Classification (1980): 01-02, 01 A 99, 70 K 10, 70 K 20, 70 K 50, 82 A 31, 92 A 08, 92 A 15

ISBN 3-540-16038-8 Springer-Verlag Berlin Heidelberg New York Tokyo
ISBN 0-387-16038-8 Springer-Verlag New York Heidelberg Berlin Tokyo

Printing and binding: Beltz Offsetdruck, Hemsbach/Bergstr.
2146/3140-543210

DEDICATION

This essay is dedicated to the memory of my friend and mentor
Elliott W. Montroll who, among other things, taught me the value
of a historical perspective in the practice of science.

PREFACE

One of my favorite quotes is from a letter of Charles Darwin (1887):

"I have long discovered that geologists never read each other's works, and
that the only object in writing a book is proof of earnestness, and that you
do not form your opinions without undergoing labour of some kind."

It is not clear if this private opinion of Darwin was one that he held to be absolutely
true, or was one of those opinions that, as with most of us, coincides with our "bad
days," but is replaced with a more optimistic view on our "good days." I hold the
sense of the statement to be true in general, but not with regard to scientists never
reading each other's work. Even if that were true however, the present essay would
still have been written as a *proof of earnestness*.

This essay outlines my personal view of how nonlinear mathematics may be of
value in formulating models outside the physical sciences. This perspective has
developed over a number of years during which time I have repeatedly been amazed at
how an "accepted" model would fail to faithfully characterize the full range of available data because of its implicit or explicit dependence on *linear* concepts. This essay
is intended to demonstrate how linear ideas have come to dominate and therefore limit
a scientist's ability to understand any given class of phenomena. In the course of this
discussion I hope to make clear how alternative nonlinear concepts encompass a much
broader range of potential applications. Thus the area of study is not biophysics, not
is it strictly biomathematics, nor is it any of the other traditional disciplines with
which we are comfortable. In part it is a step backward to a scientific attitude that
was perhaps only accepted when science was still Natural Philosophy and which did
not recognize the imposed distinctions among physics and chemistry, sociology and
anthropology, biology and ecology, etc.

A significant part of this essay is based on lectures prepared for the Integrative
Neuropsychobiology Seminar Series held by the Psychiatry Department, in the University of California, San Diego Medical School. This seminar group was started by A.
Mandell, who read earlier versions of this essay and contributed his insight regarding
both the form and substance of the material. Other people who have read various
versions of the manuscript and whose suggestions were implemented in one form or
another in the final version are: D. Criswell, R. Pathria, L. Glass, P. Russo and

S. Knapp. Also special thanks to my son Jason who wrote many of the computer codes used herein. I thank each and everyone of these people for their contributions in particular R. Rocher for her magnificent word processing of the manuscript. Of course I take full responsibility for the final form of the essay and emphasize that the implementation of my interpretation of a person's suggestion may not in fact be identical to what the person thought was said, but then that is what makes it an essay.

Finally I wish to acknowledge J. Salk for our many conversations and for introducing me to the MacArthur Foundation Mental Health Networks. The interactions with the participants of their various workshops convinced me of the timeliness of the of the present effort.

TABLE OF CONTENTS

1. INTRODUCTION

The fields of scientific knowledge[1] have been established through the painstaking effort of large numbers of dedicated investigators over long periods of time. As a scientist I am in awe of the present day rate of growth in information and in the growing number of practitioners of science. One can easily be discouraged by an inability to absorb all this information or to know all one's colleagues, or even those engaged in areas of research directly related to one's own. It appears that the modern explosion in scientific knowledge has far outstripped our collective ability to comprehend and the slower paced research of the 18th and 19th centuries has a strong appeal. This attraction results from a distortion of the true situation, however, and the feelings of isolation and impotence are not unique to the contemporary scientist. De Sola Price (1963) points out that during the past three hundred years:

" ...every doubling of the population has produced at least three doublings in

the number of scientists, so that the size of science is eight times what it was

and the number of scientists per million population has multiplied by four."

This exponential increase in the number of scientists between the 17th and 20th centuries has resulted in a similar increase in the number of books, periodicals and journals necessary to communicate and store this new found scientific knowledge, see, e.g., Figure (1.0.1). Thus 80-90 percent of the current science is a direct result of this exponential growth and consequently it must also have been *true at all times in the past* over which this growth has been present (perhaps as far back as the 17th century). De Sola Price (1963) pushes the point home by stating:

"Science has always been modern; it has always been exploding into the

population, always on the brink of its expansive revolution. Scientists have

always felt themselves to be awash in a sea of scientific literature that aug-

ments in each decade as much as in all times before."

Of course, he was aware of the fact that all growth processes eventually saturate, the growth of science being no exception, and he does discuss the changes in the perception of science and scientists as this saturation is reached. The present generation of scientists is part of this saturation region of the growth process and I maintain that this has strongly influenced the *shift in perspective* that has emerged within science in the last decade. The purpose of this essay is to indicate how this new perspective is intertwined with certain aspects of nonlinear mathematics and physics.

What is of most immediate interest to us here is the manner in which scientists have managed, and continue to manage, the ever-present flood of information referred to above. One of the dominant methods for handling this flood was/is through the

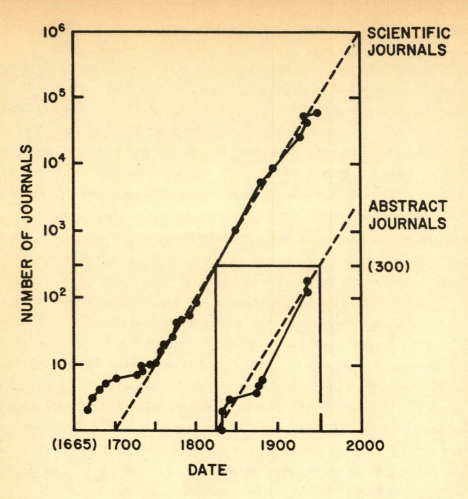

Figure 1.0.1.   The total number of scientific journals and abstract journals founded
as a function of year [from de Sola Price (1963)].

formation of new disciplines of sufficiently limited scope so as to allow mastery of the knowledge generated in the newly circumscribed area. This technique of segmenting or compartmentalizing information to manageable proportions has led to the practice of specialization and perhaps to what is now called "over-specialization." An example of this segmentation phenomenon is provided by biology. Biology began as an observational science concerned with the characteristics of the life, growth and development of organisms both as species and individuals. The experimental methods by which these characteristics were delineated were instrumental in the birth of such disciplines as physiology and histology. The theoretical explanation of large classes of data led to the theory of evolution and the acceptance of the discipline of genetics. At the interface with botany it generated ecology; at the interface with sociology and psychology it gave birth to anthropology; at the interface with physics and chemistry it sired both biophysics and microbiology. Thus biology was not isolated from the other branches of knowledge, but as any given region of overlap matured, scientists infused it with a life of its own and attempted to insulate it from its parents. This has been done under the guise of studying the newly formed area in a preliminary way through a number of simplifying assumptions. The first and foremost of these simplifying assumptions has been that the area of study can be *isolated*, even though it requires the interaction of two *previously distinct* disciplines for its conceptual formation.

In part, the discussion presented here is an attempt to uncover the hidden unity underlying the compartmentalized branches of scientific knowledge. It is an essay on a number of concepts upon which all of natural philosophy appears to be based. As an essay it is more of a personal perspective than an objective scientific paper. Its purpose is to convince the reasonable skeptic that much of what constitutes our body of *theoretical* knowledge in natural philosophy is based on linear mathematical concepts and to suggest how the more encompassing ideas of nonlinear mathematics would be better suited to the understanding of existing data sets. In particular we show by way of example that there are fundamental phenomena in all fields of natural philosophy that are intrinsically nonlinear and therefore cannot be understood by any application or extension of linear ideas. In support of this contention we also present examples in which linear ideas have actually inhibited the understanding of certain phenomena since the proper (nonlinear) interpretations violate one's intuition.

1.1 The Five Stages of Model Building

The accepted criteria for understanding a given phenomenon varies as one changes from discipline to discipline since different disciplines are at different levels of scientific maturity. In the early developmental stage of a discipline one is often satisfied with characterizing a phenomenon by means of a detailed verbal description. This stage of development reaches maturity when general concepts are introduced which tie together observations by means of one or a few basic principles, e.g., Darwin (1859) did this with evolution through the introduction of: (1) the principle of universal evolution, (2) the law of natural selection, and (3) the law of survival of the fittest. Freud (1895) did this with human behavior through the introduction of concepts such as conversion hysteria and neuroses. These investigators postulated causal relations using repeated observations of the gross properties of the systems they examined. As observational techniques became more refined additional detailed structures associated with these gross properties were uncovered. In the examples cited the genetic structure of the DNA molecule has replaced Darwin's notion of "survival of the fittest" and causal relations for social behavior are now sought at the level of biochemistry. The schism between Freud's vision of a grand psychoanalytic theory and micro-biology is no less great. The criteria for understanding in the later stages of development are quite different from those in the first stage. At these "deeper" levels the underlying principles must be universal and tied to the disciplines of mathematics, physics, and chemistry. Thus concepts such as energy and entropy appear in the discussion of micro-biological processes and are used to guide the progress of research in these areas.

An important methodological distinction emerges in the second stage of the development of a discipline. Whereas one was content with cogent verbal descriptions and qualitative arguments in stage 1, the need to quantify has been the hallmark of the later stages of development. Subsequent to the quantification of a discipline is the development of mathematical formalisms which interrelate the measurable quantities and enable the investigator to think in terms of dynamic processes and to make explicit predictions. The desire to partition the qualities of a complex biological or behavioral event into a finite number of measurable segments and then to describe the evolution of each of these segments has led to the development of a number of modeling strategies. The term model is used here to denote the mathematical representation of the biological or behavioral event along with its attendant interpretation.

The mathematical models that have been developed throughout natural philosophy have followed the paradigm of physics and chemistry. Not just in the search for basic postulates that will be universally applicable and from which one can draw deductions, but more restrictively at the operational level the techniques that have been adopted, with few exceptions, have been *linear*.

1.1.1 Linear Models

The notion of linearity arises in various guises in a number of distinct contexts as we will find out, and although all these forms are equivalent at a certain level of abstraction, they each provide a slightly different shade of meaning to the idea. Let us consider a complicated system that consists of a number of factors. One property of linearity is that the response of the action of each separate factor is proportional to its value. This is the *property of proportionality*. A second property is that the total response to an action is equal to the sum of the results of the values of the separate factors. This is the *property of independence*, see e.g., Faddeev, (1964). As a simplified example, consider a group of independent stocks and *assume* that the price of each of the stocks is directly proportional to the difference between the number of sellers owning stock and the number of buyers desiring to purchase said stock. We call this difference the buyer excess, which can in fact be negative. A positive increase of a particular buyer excess causes an increase in price which is directly proportional to that of the buyer excess. A general increase in the buyer excess distributed over the group of stocks yields an appropriate increase in price by independent amounts for each of the stocks.

Throughout the text, we shall present examples from outside the physical sciences to clarify our ideas. These examples are chosen to provide a familiar context for the explication of a concept and, hopefully, one with which the reader is comfortable. The importance of placing a new insight into a familiar setting cannot be overemphasized. It is for this reason that I avoid the expedient of keeping the discussion within the confines of physics, although most of the ideas presented herein became clear to me in this latter context. We will draw our simplified models from as many distinct branches of natural philosophy as possible and thereby avoid offending any one group of scientists more than any other.

As a second example of linearity, we choose the discipline of systems theory wherein one asserts that a process (or system) is linear if the output of an operation is directly proportional to the input. The proportionality constant is a measure of the sensitivity of the system to the input. In the preceding example, this would be the sensitivity of prices to the buyer excess. In a similar fashion the response (R) of a physical system is linear when it is directly proportional to the applied force (F). This relation can be expressed algebraically by the equation $R = \alpha F + \beta$, where α and β are constants. If there is no response in the absence of the applied force, then $\beta = 0$. A linear system is one that if two distinct forces F_1 and F_2 are applied, has the net response $S = \alpha_1 F_1 + \alpha_2 F_2$, where α_1 and α_2 are independent constants. If there are N independent applied forces denoted by the vector $\mathbf{F} = (F_1, F_2, ..., F_N)$ then the response of the system is linear if there is a vector $\boldsymbol{\alpha} = (\alpha_1, \alpha_2, ..., \alpha_N)$ of independent constant components such that $R = \boldsymbol{\alpha} \cdot \mathbf{F} = \sum_{j=1}^{N} \alpha_j F_j$. In this last equation, we clearly see that the total response of the system, here a scalar, is a sum of the independent applied forces F_j each with its own sensitivity coefficient α_j. These ideas carry over to more general systems where $\mathbf{F}$ is a generalized force vector and R is the generalized response. Consider the familiar system consisting of a student and N independent professors. Let us interpret the applied forces F_j as the prescribed homework in the jth course and α_j the level of difficulty of the material in that course as perceived by the student. Then the student's response R in the proper units may well be his anxiety level, or his self-esteem, or any of a number of other factors. From this example, we see that the response to a given set of applied forces $\mathbf{F}$ depends on the set of sensitivity coefficients $\boldsymbol{\alpha}$ which are often empirically determined.

As discussed by Lavrentév and Nikol'skii (1964), one of the most fruitful and brilliant ideas of the second half of the 1600's was the idea that the concept of a function and the geometric representation of a line are related. This connection can be realized, for example, by means of a rectangular Cartesian coordinate system. On a plane, we choose two mutually perpendicular lines (the abscissa and ordinate) on each of which we associate a positive direction. Then to each point P in the plane, we assign the pairs of numbers (x,y), which are its coordinates. With such a system of coordinates, we may represent functions graphically in the form of certain lines. Suppose we are given a function $y=f(x)$, then for each value of x the function prescribes a corresponding value of y. The graph of the function $f(x)$ is the geometric locus of points whose coordinates satisfy the equation $y=f(x)$.

Geometrically the notion of a linear relation between two quantities implies that if a graph is constructed with the ordinate denoting the values of one variable and the abscissa the values of the other variable then the relation in question appears as a straight line. In Figure (1.1.1), we depict such a curve. In systems of more than two variables, a linear relation defines a higher order "flat" surface. For example, three variables can be realized as a three dimensional coordinate space, and the linear relation defines a plane in this space. One often sees this geometrical notion employed in the analysis of data by first transforming one or more of the variables to a form in which the data is anticipated to lie on a straight line, e.g., linear regression in §2.2. Thus one often searches for a representation in which linear ideas may be valid since the analysis of linear systems is completely developed, whereas that for nonlinear

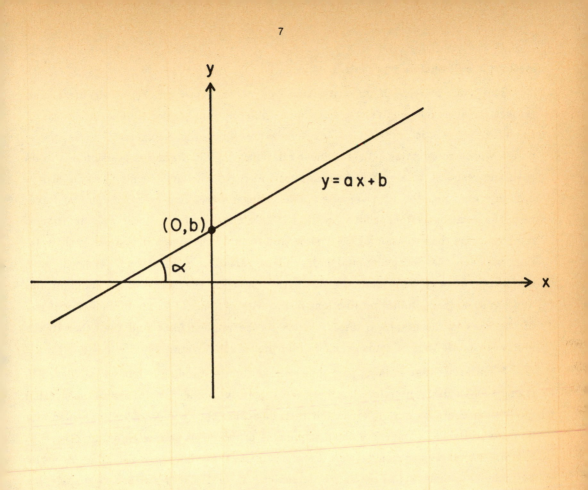

Figure 1.1.1.   The orthogonal axes define a Cartesian coordinate system. The curve $y = ax + b$ defines a straight line that intercepts the y-axis at the value $y = b$, i.e., at $x = 0$, and has a slope given by a.

systems is still relatively primative.

The two notions of linearity that we have expressed here, algebraic and geometric, although equivalent, have quite different implications. The latter use of the idea is a static[2] graph of a function expressed as the geometrical locus of the points whose coordinates satisfy a linear relationship. The former expression has to do with the response of a system to an applied force which implies that the system is a dynamic one, i.e., the physical observable changes over time even though the force-response relation is independent of time. This change of the observable in time is referred to as the evolution of the system and for only the simplest systems is the relation between the dependent and independent variables a linear one. We will have ample opportunity to explore the distinction between the above static (equilibrium) and dynamic (non-equilibrium) notions of linearity. It should be mentioned that if one of the axes in the graphical display refers to the applied force and the other to the system response then of course the two interpretations dovetail.

I will argue that in large part much of the "understanding" that was achieved in stage 1 of a given discipline has been subsequently modified not because the verbal qualitative concepts were in error, but rather because they could not be quantified in a linear world view in stage 2 of the discipline. I propose this as a working hypothesis to explain the existing schism between those scientists that view their disciplines as a global coherent whole, e.g., Freud's psychoanalytic theory, and those that view their discipline as being composed of discrete interactive elements, e.g., neuropsychobiology. In large part I think this schism exists because of an unexamined assumption made at the beginning of stage 2 development of a discipline. This assumption has to do with the ability to isolate and measure, i.e., to operationally define, a variable. In natural philosophy this operational definition of a variable becomes intertwined with the concept of linearity and therein lies the problem. To unambiguously define a variable it must be measured in isolation, i.e., in a context in which the variable is uncoupled from the remainder of the universe. This situation can often be achieved in the physical sciences (leaving quantum mechanical considerations aside), but not so for example in the social and life sciences. Thus, one must *assume* that the operational definition of a variable is sufficient for the purposes of using the concept in the formulation of a model. This assumption presumes that the interaction of the variable with other "operationally defined" variables constituting the system is sufficiently weak that for some specified conditions the interactions may be neglected. In the physical sciences one has come to call such effects "weak interactions" and has developed perturbation theories to describe successively stronger interactions between a variable and the physical system of interest (cf. §3.1). Not only is there no *a priori* reason why this should be true in the other areas of natural philosophy, but in point of fact there is a great deal of stage 1 evidence that this, in fact, is not true. We postpone the presentation of this evidence until the appropriate places in our discussion.

Note that in our discussion of stage 2 development there was no mention of the evolution of the system, i.e., no change of the system over time. One is usually content at this stage to identify the equilibrium or static relationships between concepts. As I see it the next *logical* stage of development is the specification of the dynamic relations between variables. Logic does not always dictate how scientific theories develop however, so it is quite possible that the dynamics of a process are understood before the statics. In fact in some processes the idea of a static relationship may be completely inappropriate and so stage 2 is skipped altogether. The systems approach mentioned above is an example of the stage 3 or dynamic type of modeling. At this level of abstraction the investigator attempts to manipulate the system in a controlled way in order that he/she may establish the relative values of the α's introduced earlier. These parameters may be viewed as sensitivity coefficients in one context, as correlation coefficients in a second context and so on. As sensitivity measures one would interpret $\alpha_j > \alpha_k$ to be indicative of the system being more sensitive to the *jth* stimulus than to the *kth* stimulus. As a correlation coefficient one would envision a fluctuating system whose direct response to a stimulation is unclear and therefore statistical measures, such as a correlation coefficient, must be introduced to determine (on the average) the net responsiveness of a system to a given stimulation. Note that these concepts are related and are both linear in character.

1.1.2 Nonlinear Models

These three stages of "understanding" or model building fairly well represent the traditional path of scientific maturation undergone by various disciplines. Other disciplines, such as biology and ecology, have had to push along somewhat faster in certain areas because of the pre-eminence of the nonlinear character of the systems studied. That is not to say that other disciplines such as medicine are somehow more linear, but rather that because of the abundance of certain kinds of data the biologists and ecologists had to face the problem of nonlinearity much sooner. The well known prey-predator system in which the interaction between two species leads to periodic growth and decay of the population levels necessitated the formulation of the concept of a "dynamic equilibrium." A dynamic equilibrium or "steady state" of a system is a stage 4 concept and is a generalization of the static relations of stage 2 to dynamic relations.[3] The beating of the heart and other biorhythms cannot be understood using the linear dynamic concepts from stage 3 models, however deep insight into such processes are provided by stage 4 modeling techniques. Another example of the application of the steady state concept is the notion of a *dissipative structure* in which the flux of energy (or anything else) through a thermodynamically open system results in the formation of coherent structures. This particular idea has found favor among a number of biologists, but is presented here merely as another example of the guise in which the underlying concept of a steady state in a dynamical system appears.

Another feature of nonlinear systems is the simultaneous existence of multiple steady states. The final stage of modeling development considered here is that of nonlinear dynamics and describes how the system of interest evolves towards these multiple steady states in time. The particular steady state that the system achieves asymptotically (as time becomes large) is often controlled by particular values of the parameters (control parameters) characterizing the process. In certain chemical reactions the dependence on these parameters has been explored in great detail and we will discuss some of these subsequently. The examples for stage 5 modeling are drawn from the physical sciences because that is where these ideas have been most fully developed. The application of stage 5 modeling techniques to the biological and behavioral sciences has not as yet been made (with a few notable exceptions). I believe that many such applications will suggest themselves as we examine the content of the ideas in the stage 4 and 5 description of the various phenomena.

Five Stages of Understanding

1. Detailed verbal descriptions culminating in general concepts which synthesize observations into a few fundamental principles.

2. The quantification of stage 1 concepts and their subsequent rendering into static linear (mathematical) relationships.

3. The generalization of the relationships in stage 2 into a linear dynamic description from which the relaxation of the process towards the stage 2 relations can be determined.

4. A fundamental shift in perspective to re-examine the representation of stage 1 idea to include the concept of dynamical steady states. Such states require a nonlinear representation and may have little or no direct relation to stage 3 concepts.

5. The faithful mathematical transcription of stage 1 understanding into a fully dynamical nonlinear theory whose "solutions" approach the dynamical steady states in stage 4 with increasing time.

1.2 Truth in Modeling

It has been my experience that the single greatest barrier to the effective interaction and exchange of ideas among scientists from the separate communities is the lack of an agreed upon common language. One such language is the quantitative mathematics of the physical sciences among which are the dialects of differential equations, probability theory, differential geometry, etc. Another is the qualitative mathematics of some areas of economics, e.g., game theory, or the attempts to apply the topological ideas of catastrophe theory to biology and so on. In large part, the resistance to the development of a universal language arises from the lack of agreement as to the form such a language should take, which in turn depends on the fundamental assumptions that underly how various disciplines view scientific "truth." Thus we arrive at the point of the present digression, that being to clarify in part the nature of truth in science and how our view of truth affects what we accept as an explanation.

In the life sciences, in which I arbitrarily include biology, ecology, organic chemistry, medicine, etc., the truth is very often experimental, that is to say, the truth content of a communication is associated entirely with its empirical content. This is the Lockean ideal in which a model of a system is true only to the extent that every complex proposition can be reduced to simple empirical referents that can be widely agreed upon by different observers. In §2.1 and §2.2 we examine the reliability of one's ideas about measurement, data and the certainty with which scientific concepts can be tested using data sets.

It is no surprise that communication between this empirically oriented group and those scientists that contend that truth is analytic is fraught with difficulty (if not danger). In this latter group the truth value of a model is associated entirely with its formal content. Mathematicians very often hold this point of view and maintain that the truth of a model is independent of the raw data of the external world. This is the Leibnizian ideal in which all of science must follow a predetermined path resulting in the formation of scientific "laws." For Leibniz and those of his school no amount of raw data would be sufficient to induce a general law of nature; such a law would of necessity be prior to the acquisition of data. Although I do not subscribe to this point of view there is still a certain amount of formal mathematical manipulation in this essay. These manipulations are unavoidable and in a sense are desirable, since they highlight those aspects of theory which have already been quantified.

Out of these two conflicting views of the nature of truth arose the Kantian ideal in which the truth is synthetic, that is to say, the truth does not lie separately in either empirical data or in purely rational constructs, but in both. To every theoretical construct there corresponds an empirical referent and vice versa, so that theory and data become inseparable. This is the favored view of most modern physical scientists. One is tempted to say that perhaps the latter group is somehow closer to the truth, but then we are back to the different views of truth promulgated by each of these three ideals. In our discussions we will often use the same mathematical symbol to denote both the abstract concept and the data set for the concept. This dual role for the same symbol emphasizes the synthetic nature of this viewpoint, which has been so successful in our understanding of physical phenomena. There will be other times when the data-based estimate and the theoretical value of a quantity will be distinguished in order to emphasize the preliminary nature of the concept being discussed, see e.g., §2.2.

Of course these three interpretations do not exhaust the philosophical ruminations as to the nature of truth and its role in science. Hegel maintained that truth was determined by conflict and that any complicated system is modeled simultaneously by a plan and a diametrically opposed counterplan. The purpose of the resulting conflict between the two plans is to synthesize a new plan that encompasses both the original plans. A key element of this view is that empirical data does not contain any information independent of interpretation. In particular, empirical data cannot be used to distinguish between the two plans; in fact the data can be interpreted in such a way as to be consistent with both. It should be noted that this vision of truth is not based on agreement, but rather on conflict. It is this dialectic that forms the basis of many of the recent views in economics and sociology. We stress that although dialectic conflict has been found to have explanatory value in these latter disciplines it has never been incorporated into any accepted physical theory.[4] This notion of truth will not enter into our discussions.

The final notion of truth with which we are familiar and which bears on the present discussion is that truth is pragmatic. This concept associates truth with goal orientation in a teleological sense of the system model. In this regard there is no absolute truth but only an instantaneous conformity with the goals of the model. Further, the characteristics of the modeler or rather of the experimenter must be incorporated into the goals of the model so that in a very real sense it becomes "global." This is not unlike the quantum mechanical notion that an experimenter cannot measure a physical process independently of his interaction with the process. This concept also arises in the treatment of the interaction of a quantity with the rest of the universe, i.e., the recognition of the condition that one can never truly isolate an observable. In

§2.3 we find that this non-isolatability often results in a fundamental uncertainty in what we can know about an observable and thereby limits our ability to make predictions. Thus we are often forced to make conditional or probabilistic statements about the future of a system [see §2.4], based on an uncertain knowledge of its interaction with the surroundings.

_____

[1]Scientific knowledge here refers to that body of knowledge encompassed by biology, chemistry, economics, medicine, physics, psychiatry, psychology, sociology, and so on and is not restricted to the physical sciences. We will use the old fashioned term natural philosophy to encompass all these disciplines and more.

[2]The term static means unchanging in time, as opposed to the terms dynamic or transient which denote an explicit time dependence.

[3]Here we distinguish between the traditional notion of equilibrium in which all forces are in balance, and the much more general idea of maintaining a state through the continual expenditure of energy. It is the latter picture that is more important for the biological and behavioral science.

[4]One might be able to construct a reasonable argument in which the wave-particle duality in quantum mechanics results from a dialectic conflict, but such considerations are much too far from our main purpose here and will, therefore, be abandoned.

2. ERROR ANALYSIS, STATISTICS AND OTHER UNCERTAINTIES ASSOCIATED WITH LINEARITY

Many traits of humankind have been suggested as being characteristics of the species. It seems to me that the ability and/or desire to plan for future events must be one of the dominant traits of civilized people. Such planning has certainly contributed to the survival of the species and has perhaps led to a social being with a predisposition for anticipating, if not predicting the future. In a sociological context, forecasting is important for endeavors ranging from economic to military to recreational. As scientists we strive to understand the underlying processes and to correct deficiencies in both our understanding of these processes and in the limitations of forecasting from extant data. The study of the question of our fundamental limitations of forecasting has been termed "predictability."

From one point of view, the question of predictability is philosophical; the central issue being whether the future is determined precisely and deterministically from the present or whether there is a distribution of possible futures. Until fairly recently, the epitome of the predetermined future were the particle orbits or the trajectories of rigid body motion in analytical mechanics. For example Laplace boasted that if he knew the initial positions and velocities of all the particles in the universe at one point in time then he could predict the subsequent evolution of the universe *exactly*. Physical scientists working in the more phenomenologically oriented discipline of thermodynamics knew however that the mathematical precision of analytical mechanics did not faithfully describe the apparent loss of information (organization) visible in the world around them. The tendency for the world to become disorganized and therefore less predictable received its clearest articulation in Boltzmann's presentation of the second law of thermodynamics. For the purpose here we note that the second law implies that the total entropy of a closed system will remain constant or increase over time and therefore the system either remains the same or becomes more disordered with the passage of time. Thus the ability of a scientist to predict the future state of a system from its present state decreases in time; the question of predictability being closely tied to the concept of order and organization. The representative physical system where things are totally chaotic (disorganized) is homogeneous turbulence in fluid mechanics. These two viewpoints, order and chaos, were thought to be quite distinct until the seminal work of Lorenz (1963) and the subsequent independent studies of Ruelle and Takens (1971) established that chaos (turbulence?) could arise in "deterministic" systems with as few as three degrees of freedom. A degree of freedom in physics is a variable needed to describe a dynamic process; the number of degrees of freedom is the number of such variables needed to give a complete description of a process. Thus, the point of view of one segment of the physical science community has shifted away from the vision of Laplace, and more towards the perspective that purely deterministic behavior is an *illusion* and only distributions, albeit very narrow ones in some cases, have physical significance. In the present section we develop some of the background that leads to our understanding of the role of statistics and predictability in natural philosophy. It is only much later, Section 4, that we will have developed the background necessary to understand the "order-chaos" relationship.

The mathematical and physical scientists that laid the groundwork for modern day statistics very often made their livelihood not by teaching at a University, but by calculating the odds in games of dice and cards and by casting horoscopes. The classic treatise on the theory of probability of the 18th Century was Abraham de Moivre's *The Doctrine of Chances* (1718). Most students only know de Moivre through the identity $e^{in\phi} = (\cos\phi + i\sin\phi)^n = \cos n\phi + i\sin n\phi$. It is usually not appreciated that this expression was developed to simplify certain combinatoric problems in the theory of probability. De Moivre is also interesting in that his development of statistics and probability were always directly coupled to his desire to predict. One might consider the context mundane, but the example clarifies the position of the scientist. He is the person that amidst the chaos of the world clings to the belief that there is an underlying order and a certain *predictability* and that it is his goal to enunciate the rules for this predictability.

At an intuitive level the concept of predictability is quite clear. If from a given set of circumstances, at a specified time, one can determine the configuration of these circumstances at a later time, then the later configuration is predictable from the former. One might even make the somewhat stronger statement that the later state is determined by the earlier state, so that the relationship between them is deterministic. If a set of circumstances is specified by the Ncomponent vector $\mathbf{X} = (X_1, X_2, ..., X_N)$ then in order to predict the future state of the system from its present configuration, we must specify a rule for the systems evolution. In the physical sciences the traditional strategy is to construct a set of differential equations. There equations are obtained by considering each component of the system to be a function of time, then as time changes so too do the circumstances. If in a short time interval Δt we can associate an attendant set of changes $\Delta \mathbf{X} = (\Delta X_1, ..., \Delta X_N)$ as determined by $\Delta \mathbf{X} = \mathbf{F}(\mathbf{X}, t)\Delta t$ then in the limit $\Delta t \rightarrow 0$ one would write the "equations of motion."

$$\frac{d\mathbf{X}(t)}{dt} = \mathbf{F}(\mathbf{X}, t)$$

which is a statement about the evolution of the system in time. If at time t=0 we specify the component $\mathbf{X}(0)$, i.e., the set of circumstances characterizing the system, and if $\mathbf{F}(\mathbf{X},t)$ is an analytic function of its arguments, then the evolution of the system is determined by direct integration of the equations of motion away from the initial state. This is one of the styles of thought adopted from the physical sciences into the biological and behavioral sciences. The use of this modeling technique is often attributed to Newton because of his introduction of the three laws of motion into mechanics; however, nowhere in the *Principia* will one find the famous equation $\mathbf{F} = d\mathbf{p}/dt$ where $\mathbf{F}$ is the force on a particle of momentum $\mathbf{p}$. This use of differential equations as the starting point for the description of particle and continuum mechanics was initiated and populatized by the Bernoulli clan, (Jacob, John, Nicolis and Daniel) along with their illustrious student L. Euler, see e.g., Montroll and Schlesinger (1984) for some historical comments.

The mathematicians have categorized the solutions to such equations for the simplest kinds of systems. One way to describe such systems is by means of geometric constructions in which the solution to an equation of the above form is depicted by a curve. The coordinate axes necessary for such a construction are the continuum of values that the vector $\mathbf{X}$ can assume, each axis being associated with one component of the vector $\mathbf{X}$. Consider a space having axes labeled by the components of the dynamical system. In Figure (2.0.1.) such a space is given for $\mathbf{x} = (x_1,x_2)$ and is called a phase space. A point in the phase space gives a complete characterization of the dynamical system at a point in time. This is because a single point in an N-dimensional phase space is an N-triple of values and it is this N-triple of values that characterize the state of the system. In Figure (2.0.1.) we depict the basic topological forms of the solutions to two-dimensional linear dynamical equations. A trajectory or orbit in this space traces out the evolution of the dynamical system; time is a parameter which indexes each point along the curve. The field of trajectories initiated from a set of initial conditions is often referred to as the flow field. If for example the flow field asymptotically $(t \rightarrow \infty)$ converges to a single point in phase space, this is called a fixed point [cf. focus]. If the flow field converges to a single closed curve this is called a limit cycle [cf. Figure 2.0.2.]. All this behavior is deterministic and much of it is understood, that is to say, the motion is predictable, see e.g., Arnold (1981).

We may then (in one sense) associate predictability with the stability of orbits in phase space, i.e., if a small change in the orbit at a point does not change the properties of the system it is said to be *stable*, if such a small change yields a large deflection from its previous behavior the orbit is called *unsable*. Thus a stable system is one for

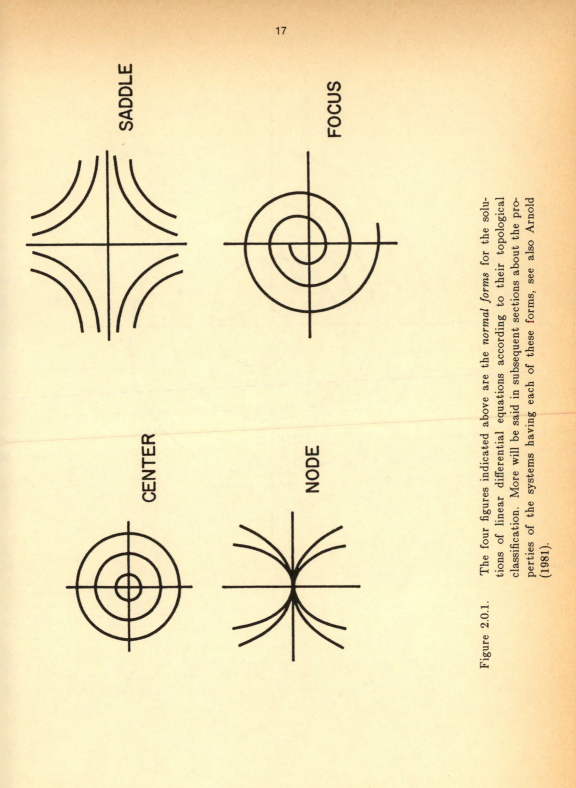

Figure 2.0.1. The four figures indicated above are the *normal forms* for the solutions of linear differential equations according to their topological classification. More will be said in subsequent sections about the properties of the systems having each of these forms, see also Arnold (1981).

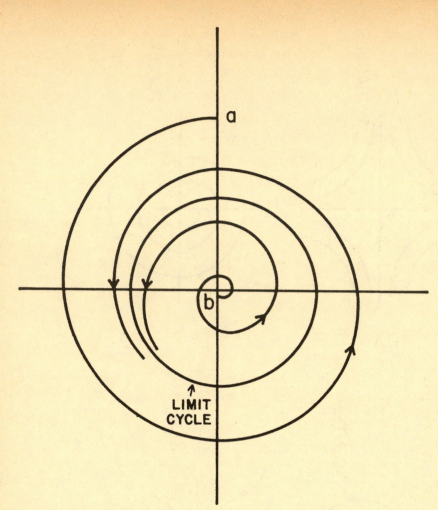

Figure 2.0.2.　The two points *a* and *b* in the figure are possible initial conditions for the system. When the system can manifest limit cycle behavior the orbits approach this cycle asymptotically and lose all memory of their initial state.

which two orbits that are nearby initially remain nearby throughout the evolution of the system. An unstable system is one for which two orbits that are nearby initially become arbitrarily distant one from the other as time increases. This sensitivity to the initial state in an unstable system is manifest in an attendant lack of predictability. One cannot predict a given final state for a system if a slight change in its present state leads to such uncontrollable changes in the way the system evolves. By way of contract consider a pencil balanced on its point, we know this system is locally unstable, but there is a point of global stability in which the pencil is at rest on the table. All perturbations away from the unstable point go to this state of equilibrium. This kind of stability is a static property of the system and requires the forces acting on the pencil to be in balance. The kind of stability involving orbits in phase space is dynamic and can (and does) arise in systems in which forces are unbalanced so that there is a flux through the system. These concepts will be fully discussed subsequently. For the moment we merely note that this is a rather abstract formulation of the difficulties encountered in interpreting and making observations.

A less sophisticated uncertainly is related to the problems of measurement and the errors resulting therefrom and is discussed in §2.1. The *calculus of errors* has to do with the precision of numerical data and the results of calculation, not with errors that are a consequence of false mathematical reasoning nor necessarily with questions of local or global stability. Errors can be either *systematic*, for example those associated with the measuring instrument, or they can be random, for example those associated with the observer. The former errors may in principle be eliminated from the observations, but the latter are uncontrollable random influences that arise during the measuring process and constitute a fundamental uncertainty in the data record.

It will become clear that not only are the traditional methods for handling data, such as regression analysis, based on linear concepts (even when purported to be nonlinear), but the whole idea of additive fluctuations (errors) in a dynamic system is predicated on linearity. In §2.2 we indicate how the ideas proposed by Gauss (1809) for the development of the "law of errors," discussed in §2.1, are identical to those reinvented a century later for mathematical statistics; in particular for the linear regression of a function to a data set. A physical model is presented in §2.3 to support the point of view that it is virtually impossible to obtain an error-free measurement of a variable due to the coupling of that variable to the rest of the universe. These three sections taken together constitute a certain basic understanding of how errors, fluctuations and stochastic processes are viewed in natural philosophy.

In §2.4 we examine the concept of predictability, or rather uncertainty, as it is manifest through the use of a probability distribution. Just as the dynamic variables describing a system change in time to mimic the system's evolution, so too does the probability evolve to describe the most likely future behavior of the system. The relation between the dynamic description including fluctuations discussed in §2.3 and the evolution equation for the probability density is presented here.

2.1 Distribution of Errors

In our earlier comments we mentioned the notion of the operational definition of a variable, that is to say, a variable defined by how it is measured. However one learns in their first excursions into the experimental sciences that no measurement is absolutely reproducible. Each time one measures a given quantity a certain amount of estimation is required, e.g., in reading the markings on a ruler or a pointer on some gauge, so that if one measures a quantity x a given number of times, N say, then instead of having a single quantity X one instead has a collection of quantities X_1, X_2, $\cdots X_N$. Such a collection is often referred to as an ensemble. T. Simpson (1755) was the first scientist to recommend that all the measurements taken in an experiment be utilized in the determination of a quantity and not just those considered to be the most reliable, as was the custom of the time. He was the first to recognize that the observed discrepancies between results follow a pattern that are characteristic of the ensemble of measurements. His observations were the forerunner of the *law of error* which asserts that there exists a relationship between the magnitude of an error and its frequency of occurrence in an ensemble. If we consider the relative number of times an error of a given magnitude occurs in a population of a given (large) size, i.e., the frequency of occurrence, we obtain an *estimate* of the probability an error of this order will occur. From the above ensemble we can define an average value, denoted by a line over the variable:

$$\overline{X} = \frac{1}{N} \sum_{j=1}^{N} X_j \ .$$

(2.1.1)

The mean value $\overline{X}$ is often thought to be an adequate characterization of the measurement and thus an operational definition of X is associated with $\overline{X}$. Simpson was also the first to suggest that the mean value be accepted as the best value of the measured quantity. He further proposed that an isosceles triangle be used to represent the theoretical distribution in the measurements around the mean value. Subsequently, it was recognized that to be more quantitative one should examine the degree of variation of the *measured value* away from its average value. The magnitude of this variation is defined by the standard deviation (σ) or variance (σ^2) of the measurements:

$$\sigma^2 \equiv \frac{1}{N} \sum_{j=1}^{N} (X_j - \overline{X})^2 \tag{2.1.2}$$

which using (2.1.1) can be reduced to

$$\sigma^2 = \overline{X^2} - \overline{X}^2 \; . \tag{2.1.3}$$

[See e.g., Beers (1953)].

In the continuous limit, i.e., the limit in which the number of independent observations of a quantity approaches infinity, the characteristics of any well-behaved measured quantity are specified by means of a distribution function. The general concept is that any particular measurement has little or no meaning in itself, it is only the collection of measurements, i.e., the ensemble, that has a physical interpretation and this meaning is manifest through the distribution function. The distribution function, also called the probability density, associates a probability with the occurrence of an event in the neighborhood of a given magnitude. For example in Figure (2.1.1) we indicate the frequency of occurrence of adult males of a given height in the general population of the British Isles. From this distribution it is clear that the likelihood of encountering a male six feet in height on your trip to Britain is substantial, that of meeting someone six feet six inches tall is much less so, the probability of seeing someone greater than ten feet tall is zero. Quantitatively, the probability of meeting someone with a height X in the interval $(x, x + \Delta x)$ is $P(x)\Delta x$ where $P(x)$ is the desired distribution function. The solid curve in Figure (2.1.1) is given by a mathematical expression for the functional form of $P(x)$. Such bell-shaped curves whether they are from measurements of heights or of errors are described by the well known distribution of Gauss which we now discuss.

Gauss (1809) was the first scientist to systematically investigate the properties of measurement errors and in so doing set the course of experimental science for over a century. He postulated that if each observation in a sequence of measurements was truly independent of each of the other measurements then $\xi_j = X_j - \overline{X}$ is a random variable and ξ_j and ξ_k are statistically independent for $j \neq k$. If one defines ξ_j as the error in the jth measurement then the mean error is zero by construction and the standard deviation (2.1.2) is given by

$$\sigma^2 = \frac{1}{N} \sum_{j=1}^{N} \xi_j^2 \; . \tag{2.1.4}$$

Gauss noted that the probability I of obtaining a value in the interval $(\xi, \xi + \Delta \xi)$ in any given measurement where $\xi = X - \overline{X}$, is given by $P(\xi)\Delta\xi$:

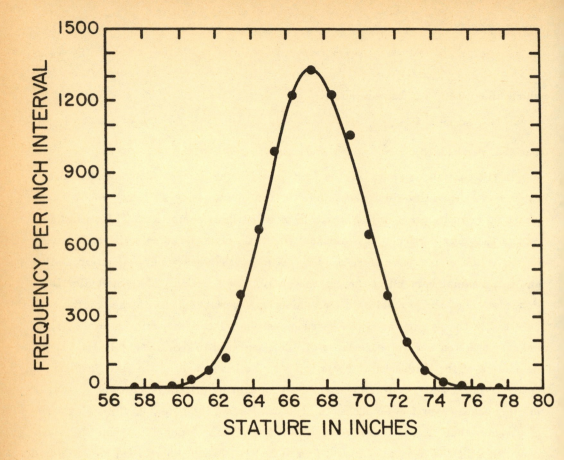

Figure 2.1.1.   The distribution of the height for adult males in the British Isles [taken from Yule and Kendall (1950)].

$$I = P(\xi)\Delta\xi \quad . \tag{2.1.5}$$

In a sequence of N measurements we assume that we have obtained the values $X_1, X_2, ..., X_N$ with deviations $\xi_1,....,\xi_N$, respectively. If we segment the range of values of ξ into N intervals each of width $\Delta\xi$ and such that

$$P(\xi_j)\Delta\xi = \text{probability of observing the deviation } \xi_j \quad , \tag{2.1.6}$$

then the probability of observing this combination of deviations is

$$I = \prod_{j=1}^{N} P(\xi_j)\Delta\xi = P(\xi_1)....P(\xi_N)(\Delta\xi)^N \quad . \tag{2.1.7}$$

According to Gauss (1809), the estimation of the value for X appears *plausible* if the set of measurements $X_1,..., X_N$ resulting in $\bar{X}$ is the most *probable*. Thus X is determined in such a way that the probability I is a maximum for $X = \bar{X}$. To determine this form of the probability density we therefore impose the condition

$$\frac{d}{dX} \ln I = \sum_{j=1}^{N} \frac{d\xi_j}{dX} \frac{\partial}{\partial\xi_j} \ln I$$

$$= - \sum_{j=1}^{N} \frac{\partial}{\partial\xi_j} \ln P(\xi_j) = 0 \tag{2.1.8}$$

where we have used the fact that $d\xi_j/dX = -1$ for all j. The constraint (2.1.8) is the mathematical rendition of the desirability of having $\bar{X}$ as the most probable value of X.

We now assume that the $j\,th$ logarithmic derivative can be expanded as a polynomial in ξ_j, i.e.,

$$\frac{\partial}{\partial\xi_j} \ln P(\xi_j) = \sum_{l=0}^{\infty} b_l \, \xi_j^l \tag{2.1.9}$$

where the constants b_l are determined by substituting the series expression, (2.1.9) into (2.1.8). The condition (2.1.8) is satisfied if all the b_l are zero except $l=1$; since $\sum_{j=1} \xi_j = 0$ this coefficient need not be zero to satisfy the constraint. Thus we have

$$\frac{\partial}{\partial\xi_j} \ln P(\xi_j) = b_1\xi_j \quad . \tag{2.1.10}$$

which integrates to

$$P(\xi_j) \propto \exp\left\{\frac{1}{2} b_1 \, \xi_j^2\right\} \quad . \tag{2.1.11}$$

Now since $P(\xi)$ is maximum at $\xi = 0$, i.e., for $X = \bar{X}$ as Gauss required, and decreases symmetrically to zero on either side, we choose b_1 such that $b_1 < 0$ and

$$\int_{}^{\infty} P(\xi)d\xi = 1 \quad ; \quad \int_{}^{\infty} \xi^2 P(\xi)d\xi = \sigma^2 \tag{2.1.12}$$

yielding the normalized probability density

$$P(\xi_j) = \frac{1}{\sqrt{2\pi}\,\sigma_j}\ \exp\left\{-\xi_j^2/2\sigma_j^2\right\}\ .$$

(2.1.13)

Equation (2.1.13) is the normal or Gauss distribution of error and using the multiple convolution relation[1] of Gauss distributions yields

$$P(\xi) = \frac{1}{\sqrt{2\pi}\,\sigma}\ \exp\left\{-\xi^2/2\sigma^2\right\}$$

(2.1.14)

with

$$\sigma^2 = \sum_{j=1}^{N} \sigma_j^2\ .$$

(2.1.15)

In Figure (2.1.2) the probability density is graphed as a function of the measured quantity ξ. In the discussion offset from the main text we have argued that the distribution of error is given by the Gauss distribution:

$$P(x) = \frac{1}{\sqrt{2\pi}\sigma}\ \exp\left\{-(x-\bar{x})^2/2\sigma^2\right\}$$

(2.1.16)

from which it is clear that the distribution is completely characterized by the mean $\bar{x}$ and variance σ^2. The probability of a measurement having a small error is substantial, i.e., having a value in the neighborhood of $\bar{x}$ is quite likely, whereas the probability of having a large error is quite small, i.e., values of x greatly different from $\bar{x}$ are unlikely. Thus we can replace the definition of the mean value (2.1.1) and variance (2.1.2) by using the distribution (2.1.16) itself. This distribution describes the statistics of the ensemble of measurements so that under the conditions specified for the validity of (2.1.16) we can replace (2.1.1) with

$$\bar{x} = \int_{-\infty}^{\infty} xP(x)dx$$

(2.1.1')

and (2.1.2) with

$$\sigma^2 = \int_{-\infty}^{\infty} (x-\bar{x})^2\,P(x)\,dx$$

(2.1.2')

so that the mean and variance are expressed as weighted averages over the ensemble distribution function. Gauss (1809) actually phrased his argument in the form of a postulate, that being if the mean $\bar{x}$ of a sample of an ensemble of measurements $\{X_j\}$ is the most probable estimate of the measured quantity, then the ensemble has a normal distribution function.

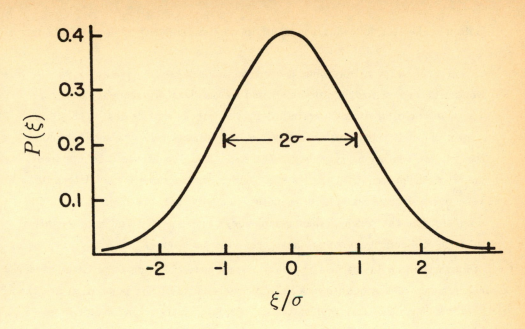

Figure 2.1.2.   The probability density $P(\xi)$ given by the Gauss distribution (2.1.14) is graphed as a function of the normalized variable ξ/σ.

Galton makes the following comments as to the regularity of form of the law of errors:

"I know of scarcely anything so apt to impress the imagination as the wonderful form of cosmic order expressed by the 'Law of Frequency of Error.' The law would have been personified by the Greeks and deified, if they had known of it. It reigns with serenity and in complete self-effacement, amidst the wildest confusion. The huger the mob, and the greater the apparent anarchy, the more perfect is its sway. It is the supreme law of Unreason. Whenever a large sample of chaotic elements are taken in hand and marshalled in the order of their magnitude, an unexpected and most beautiful form of regularity proves to have been latent all along."

Even though the individuals in a population vary, here the population is the sequence of measurements of the quantity X, the characteristics of the population are themselves very stable. The fact that statistical stability emerges out of individual variability has the appearance of order emanating from chaos and has inspired a number of metaphysical speculations. Aside from the above quote from Galton we will not follow up on any of these speculations. [See e.g., T.C. Tippett (1956).]

The proof that a sum of measurements converges to a variable whose statistics are given by a Gauss distribution relies on a number of important assumptions: (1) each error in the measurement process is statistically independent of any of the other errors; (2) the errors in the measurement process are additive; (3) the distribution describing each of the random errors ξ_j, $j=1, \cdots, N$ is the same and finally, (4) the variance of the distribution is finite. With these assumptions the *Central Limit Theorem* can be used to obtain the Gauss distribution. As Poincaré (1913) quotes apropos of the law of errors:

"All the world believes it firmly, because the mathematicians imagine that it is a fact of observation and the observers that it is a theorem of mathematics."

He was of course alluding to the fact that nature does not necessarily satisfy conditions (1)-(4) in a given sequence of measurements and one must examine separately each process that is measured.

We have based our introduction to the law of error on the notion that the Central Limit Theorem is applicable to a sum of identically distributed random variables. The reader who is unfamiliar with these mathematical concepts can gain some experience with them by using simple computer programs. In the Appendix we provide a computer program written in Basic (GAUSS RANDOM NUMBERS) to generate random numbers whose statistics are approximately given by a Gauss distribution. The distribution is approximate because the generated ensemble of measurements is finite.[2] The computer generated ensemble is constructed by adding a sequence of 12 random numbers using the random number generator included as part of the hardware of most personal computers. This number (12) is sufficiently large that the normalized sum defines a Gauss random number. A collection of N1 such realization constitutes our ensemble. These realizations are put in increasing order and the relative number of "measurements" occuring in a given interval $(x, x + \Delta x)$ is calculated. The relative frequency in each interval constitutes a "histogram" which is a discrete representation of a distribution function. The program is simple enough to run and by varying the prescribed mean (xav), variance (var) and the number of realizations in our ensemble (N1) one can gain an appreciation for the variability in "real" data sets.

A useful way to model the law of errors developed by Gauss is by means of an object participating in a random flight or random walk. The random walk model provides a physical picture of how one can construct a sum of random variables satisfying conditions (1) through (4) above. Its utility lies in being able to extend these ideas beyond the limited context of error analysis as well as being able to relax certain of the above constraints to obtain distributions other than that of Gauss. [See e.g., Chandrasekhar (1943), Wang and Uhlenbeck (1945) and for a more recent review Montroll and West (1979) or Montroll and Shlesinger (1984).]

Random walks had their birth under that name due to the following inquiry on "The Problem of the Random Walker" by the biostatistician Karl Pearson (1905) in the journal *Nature*:

"A man start from a point 0 and walks l yards in a straight line: he then turns through any angle whatever and walks another l yards in a second straight line. He repeats this process n times. I require the probability that after these n stretches he is at a distance between r and $r + \delta r$ from his starting-point 0. The problem is one of considerable interest, but I have only succeeded in obtaining an integrated solution for two stretches. I think, however, that a solution ought to be found, if only in the form of a series of powers of $1/n$, where n is large."*

Lord Rayleigh responded in the same issue of *Nature* with the following:

"This problem, proposed by Prof. Karl Pearson in the current number of *Nature,* * is the same as that of the composition of n iso-periodic vibrations of unit amplitude and of phases distributed at random, considered in *Philosophical Magazine*, x. p. 73, 1880; XLVII. p. 246, 1899 (*Scientific Papers, I. p.491;*

IV. p. 370). If n be very great, the probability sought is

$$2n^{-1} e^{-r^2/n} r dr \quad .$$

Probably methods similar to those employed in the papers referred to would avail for the development of an approximate expression applicable when n is only moderately great."

Note that the expression given by Rayleigh is the two-dimensional generalization of (2.1.21) given below.

Let us consider a random walker who at each step on the one-dimensional lattice of Figure (2.1.3) steps to the right with a probability p and to the left with a probability $q = 1-p$. The coefficient p of $e^{i\phi}$ in the expression

$$\left(p \, e^{i\phi} + qe^{-i\phi} \right)$$

represents the probability of the first step being to the right while the coefficient q of $e^{-i\phi}$ represents the probability of the first step being to the left. The coefficient p^2 of $e^{2i\phi}$ in

$$\left(p \, e^{i\phi} + qe^{-i\phi} \right)^2$$

represent the probability of a walker being at $s=2$ after two steps, the coefficient $2pq$ of $e^{io\phi}$ the probability that he has returned to the origin, and the coefficient q^2 of $e^{-2i\phi}$ the probability that he ends at $s=-2$ after two steps. Generally

$$P_n(s) = \text{probability that walker is at s after n steps}$$

$$= \text{coefficients of } e^{is\phi} \text{ in } \left(pe^{i\phi} + qe^{-i\phi} \right)^n$$

$$= \frac{1}{2\pi} \int_{-\pi}^{\pi} \left(p \, e^{i\phi} + q \, e^{-i\phi} \right)^n e^{-is\phi} \, d\phi \tag{2.1.17}$$

since the identity

$$\frac{1}{2\pi} \int_{-\pi}^{\pi} e^{-im\phi} \, d\phi = \delta_{m,0}$$

implies that the integral operator

$$\frac{1}{2\pi} \int_{-\pi}^{\pi} e^{-is\phi} \, d\phi$$

filters out the coefficient of $e^{is\phi}$ in the Fourier series representation of a periodic function of period 2π. In the symmetrical case with $p=q=1/2$

$$P_n(s) = \frac{1}{2\pi} \int_{-\pi}^{\pi} (\cos\phi)^n e^{-i\phi s} \, d\phi = \frac{n!}{\left[\frac{1}{2}(n+s) \right]! \, \left[\frac{1}{2}(n-s) \right]!} \left(\frac{1}{2} \right)^n \tag{2.1.18}$$

When n is even the only possible values of s are even with $|s| \leq n$ and when n is odd s is odd also with $|s| \leq n$.

The above integral (2.1.18) can also be written as

$$P_n(s) = \frac{1}{2\pi} \int_{-\pi}^{\pi} (\cos\phi)^n \, e^{-i\phi s} \, d\phi$$

$$= \frac{1}{2\pi} \int_{-\pi}^{\pi} \exp\left\{ -i\phi s - n\left[\frac{1}{2} \phi^2 + \frac{1}{12} \phi^4 + \dots \right] \right\} d\phi \quad ; \tag{2.1.19}$$

by setting $x = \phi n^{1/2}$ (2.1.19) becomes

$$P_n(s) = \frac{1}{(4n\pi^2)^{1/2}} \int_{-\pi\sqrt{n}}^{\pi\sqrt{n}} \exp\left\{ -ix \frac{s}{\sqrt{n}} - \frac{1}{2} x^2 - \frac{x^4}{12n} + \cdots \right\} dx \tag{2.1.20}$$

Hence as $n \to \infty$ one obtains the Gaussian form for $P_n(s)$

$$P_n(s) \sim \frac{1}{(4\pi^2 n)^{1/2}} \int_{-\infty}^{\infty} \exp\left\{ -ix\frac{s}{\sqrt{n}} - \frac{1}{2} x^2 \right\} dx$$

$$\sim \frac{1}{\sqrt{2\pi n}} \, e^{-s^2/2n} \quad . \tag{2.1.21}$$

Now suppose that a is the lattice spacing in our one-dimensional lattice and that τ is the time interval between steps. Then, with $x = sa =$ total displacement in time $t = n\tau$, this distribution is related to the probability of a displacement in the interval $(x, x + dx)$ in time t through

$$P_n(s)ds = P(x,t)dx , \quad ds = 1$$

so that,

$$P(x, t) = \frac{1}{(4\pi Dt)^{1/2}} \exp\left\{ -x^2/4Dt \right\} \tag{2.1.22}$$

with $D = a^2/2\tau$. Thus we see that the Gauss form of the law of errors can be understood using random walk arguments.

One of the more important elements of research science is the cultivation of an intuition about the process under investigation. In the present context this implies that one be able to appreciate the physical implications of an equation such as (2.1.22) from different perspectives. To assist the reader in the development of this intuition we have listed a program (RANDOM WALK) in the Appendix that numerically simulates the above random walk process. This program generates the random walk in real time, each step taking a unit time to complete. In Figure (2.1.4a) we show a typical result of the calculation, the steps generated by the random walker on a two

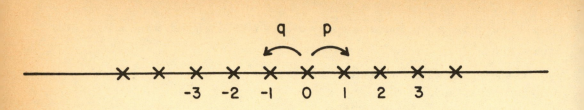

Figure 2.1.3. A discrete one-dimensional lattice is depicted on which a random walker steps to the right with probability p and to the left with probability $q = 1-p$.

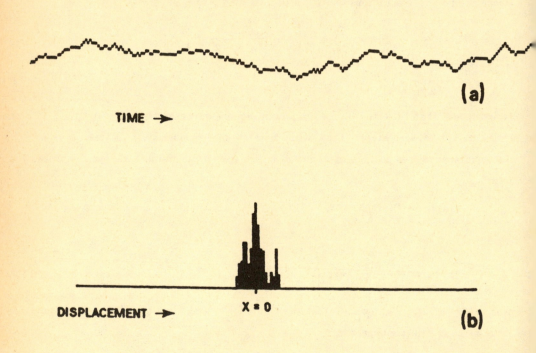

Figure 2.1.4. The steps taken by a random walker along a one-dimensional line are indicated in (a). The histogram counting the number of times the walker has occupied a given position along the x-axis during his walk is depicted in (b).

dimensional plane. In this figure we see a splatter of points more dense in the region of the origin, where the walk was initiated, than elsewhere. The computer code wil enable the reader to see the order in which these points are visited by the walker., In Figure (2.1.4b) the distribution of sites visited by the random walker is give along the x-axis. This restriction to one-dimension has been imposed only for the purpose of visualization, so that one can compare the analytic expression (2.1.22) with the histogram generated by the random walker.

It is interesting to note that this style of thought was developed in a probability context by de Moivre (1732), but the application of his result to the theory of error required another century to pass. The application of the binomial distribution (2.1.18) to considerations of the law of errors was made by Hagen (1837). His arguments although somewhat different in algebraic detail were concerned with the development of the "normal" or Gauss distribution of errors starting from a discrete distribution of the form (2.1.18). Romanowski (1979) points out that Haken presents the foundation for a theory of the law of errors in the form of a single compact "theorem." Romanowski partitions Hagen's assumptions into the set:

(a)   each measurement is distributed by an "infinite" number of causes of error

(b)   each cause of error produces an error which can be either positive or negative ("elementary error")

(c)   all elementary errors have the same absolute value

(d)   the probability p that an elementary error will be positive is equal to the probability q that it will be negative.

He (Romanowski) further states:

"In the development of the science of measurements, Hagen's work played the role of a handbook to several generations of scientists and engineers. Its fame did not equal that of other great works, e.g., of Gauss' *Theoria motus*; this was perhaps due (at least partly) to Hagen's attitude: he did not emphasize sufficiently the originality of his own procedures but only the fact that they confirm the formulae established by Gauss."

For our present purposes we focus our attention on the assumption of linearity made in the derivation of the Gauss distribution. If this assumption were violated then one could *not* construct the sum used in the Central Limit Theorem argument nor the filtering process used in the random walk discussion. Many examples from natural philosophy can in fact be found in which this assumption is violated. Consider the distribution of IQ's depicted in Figure (2.1.5). The expected number of people having IQ's in various intervals seems to depend partly on genetic factors and partly on environmental factors. From the figure we see that the distribution falls more steeply on the high side of the IQ than on the low side, perhaps indicating a constraint at the high end that is not present at the low. This distribution is clearly *not* of the Gauss form so that whatever the proper explanation for the form of the distributon, it does not result from a *linear* superposition of independant factors. A more severe deviation from the Gauss form is indicated in Figure (2.1.6) where the percent of authors publishing exactly n papers is plotted as a function of n. The straight line is Lotka's law which is an exact inverse-square law [Lotka (1926)]. The interactive mechanisms underlying such power law distributions is however quite different from those necessary for the Central Limit Theorem argument and are taken up in §3.2 where nonlinear mechanisms in fluctuating proceses are discussed.

Consider for the time being a complex task that consists of a number of subtasks, n say, each of which must be successfully completed before the overall task is accomplished. If P is the probability of completion of the complex task and p_j is the probability that the j th subtask is completed, then by the product rule for joint probabilities the total probability can be written as

$$P = \prod_{j=1}^{N} p_j = p_1 p_2 \ldots p_N \ .$$

(2.1.23)

One can replace (2.1.23) by a sum of variables by taking the logarithm of the product to obtain

$$\log P = \sum_{j=1}^{N} \log p_j \ .$$

(2.1.24)

Since by assumption the p_j are statistically independent random variables, so are the $\log p_j's$. If the second moment of $\log p_j$ is finite and n is large then by the Central Limit Theorem $\log P$ has a Gauss distribution.

As a concrete example of the occurrence of a log-normal distribution let us again consider the distribution in the number of research papers published by a scientist. There are a number of contingencies which must be met before a paper is published: one first needs to have an idea in the form of a research problem (p_1), then one needs the technical ability to solve the problem (p_2), then one needs the time and money to do the research (p_3), ... and so on. W. Shockley (1957) was the first to give this argument for research papers and using the publication records of scientists at the Brookhaven National laboratories found that indeed one does obtain a log-normal

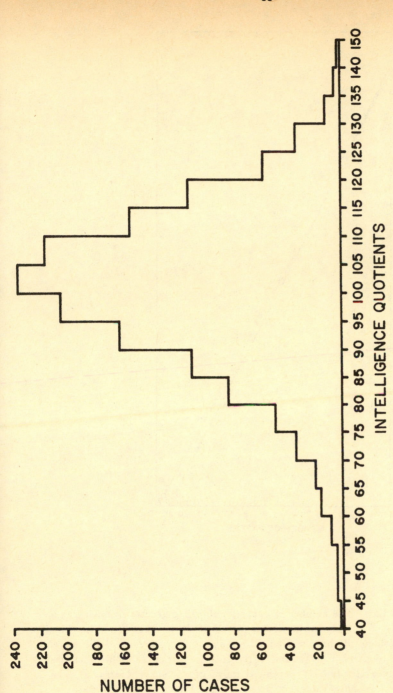

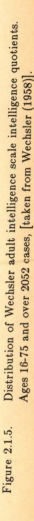

Figure 2.1.5. Distribution of Wechsler adult intelligence scale intelligence quotients. Ages 16-75 and over 2052 cases, [taken from Wechsler (1958)].

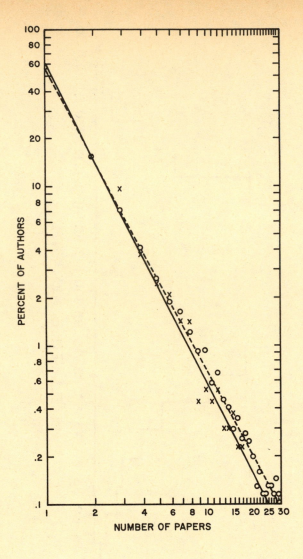

Figure 2.1.6.   Lotka's Law. The number of authors publishing exactly *n* papers, as a function of *n*. The open circles represent data taken from the first index volume of the abridged *Philosophical Transactions of the Royal Society of London* (17th and early 15th centuries), the filled circles those from the 1907-16 decennial index of *Chemical Abstracts*. The straight line shows the exact inverse-square law of Lotka. All data are reduced to a basis of exactly 100 authors publishing but a single paper.

distribution [cf. Figure 2.1.7]. Other fields in which log-normal distributions have been observed are income distributions, body weights, sound measurements (in decibels), rainfall,etc. We will discuss a number of these applications in the sequel. Also we will address the question of how the log-normal distribution is related to the power-law distribution previously discussed.

We emphasize that the log-normal distribution arises from a nonlinear transformation of the original process. As a matter of fact F. Galton motivated the introduction of the log-normal distribution into statistics by means of the observation that certain categories of events are better classified through geometric averages than through arithmetic averages, i.e., through roots of products rather than through simple linear sums.

Of course the Gauss and log-normal do not exhaust the number of possible distributions that characterize processes of interest. There are many others, a large class of which have been obtained recently using modern scaling theories. They are mentioned here in order to emphasize that although there is a large class of distributions, the *stable distributions*[3] that rely on linearity, it is only that due to Gauss which has a finite mean and variance. We will find in subsequent sections that the random walk approach to modeling will enable us to gain insight into certain of these non-Gauss processes, *in particular* those that are stable. Further, we will see how stable and power-law distributions are related, but not before we have discussed the reasons for their existence. The latter has to do with nonlinear interactions.

2.2 Data Analysis

The distribution of errors would not be of such importance to our discussion if it had not had such a profound influence on the development of concepts used to interpret data in Natural Philosophy over the last century. The concepts of regression analysis, maximum likelihood estimates and the method of least squares (all to be described below) can each in its turn be traced back to Gauss and his study of the calculus of errors. However, a clear distinction must be made between the theory of errors and mathematical statistics, since the latter evolved almost completely independently of the former even though the formal content of the two theories is essentially identical (Seal, 1967). It is quite likely that the prior existence of these ideas in the lore and literature of the physical and engineering sciences had an influence, even if indirectly, on the strategies developed in mathematical statistics. Of course it would require an exhaustive historical study to document such a connection. One such overlap we have already noted with regard to Pearson's inquiry and Rayleigh's response on random walks.

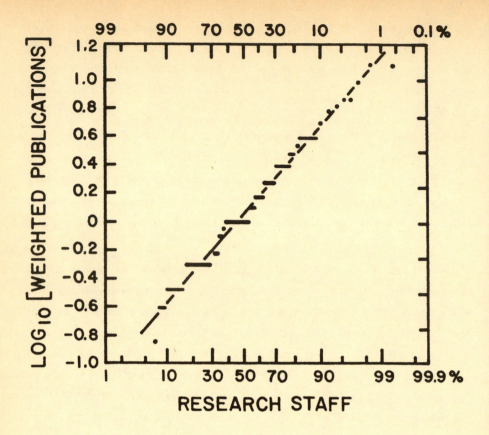

Figure 2.1.7. Cumulative distribution of logarithm of "weighted" rate of publication of Brookhaven National Laboratory for 88 members of the research staff expressed as percent of 95, plotted on probability paper [W. Shockley (1957)].

2.2.1 Discrete Data and Linear Regression

The notion of an error has direct appeal to the physical scientist since that person assumes that there is a *correct* number associated with the quantity of interest and that the variations in the measurements are deviations from this exact number due to random causes. Scientists make repeated measurements of the same object in order to form the collection of measurements, i.e., the ensemble of errors,[4] from which the statistical properties of the process can be determined. By contrast, a biologist studying the evolution of, say an animal population, concerns himself with a large number of similar but independent members of the population. Similarly a psychologist or sociologist examines the variability of independent responses chosen by respondents from a finite number of choices on a questionnaire. The variability in measurements in these latter two examples is the proper subject of mathematical statistics and on the surface does not bear a direct relation to the calculus of errors.

For the moment let us not distinguish between the above interpretations and merely assume that we have two measurable quantities x and y that are *theoretically* related by $y = F(x)$, i.e., the function F associates a unique value of y with each value of x. In a number of instances the detailed form of the functional relation is not of interest; it being sufficient to expand the function $F(x)$ in a Taylor series about a selected point $x = x_0$:

$$y = F(x_0) + (x-x_0)F'(x_0) + \frac{1}{2}(x-x_0)^2 F''(x_0) + \dots \qquad (2.2.1)$$

The Taylor expansion of a function $F(x)$ about a point $x = x_0$ can be understood using the equivalence between graphical and algebraic expressisons. In Figure (2.2.1a) we draw the curve $y = F(x)$ and consider the arbitrary but definite point on the curve $x = x_0$. The *first approximation* to the entire function is the *value* $F^{(1)}(x) = F(x_0)$, which would be exact *if* $F(x)$ were the horizontal line given in Figure (2.2.1b). The *second approximation* to the entire function is given by the *slope* of $F(x)$ at $x = x_0$ which is the change in the value of the function (ΔF) over a small interval Δx centered on $x = x_0$. Thus to first order the value of the function is $F^{(2)}(x) = F(x_0) + \frac{\Delta F}{\Delta x} \mid_{x=x_0}(x-x_0)$ which is indicated in Figure (2.2.1c). This approximation to the function looks like the actual function in the immediate vicinity of $x=x_0$, but deviates strongly from it as one gets further from the point of construction. The *third approximation* to the entire function is given by the *curvature* of $F(x)$ at $x = x_0$ which is the change in the slope of the function $\Delta\left(\frac{\Delta F}{\Delta x}\right)$ over a small interval Δx centered on $x = x_0$. Thus to second order the value of the function is $F^{(3)}(x)=F(x_0) + \frac{\Delta F}{\Delta x} \mid_{x=x_0}(x-x_0) + \frac{1}{2}\frac{\Delta^2 F}{\Delta x^2}\mid_{x=x_0}(x-x_0)^2$ which is indicated in Figure

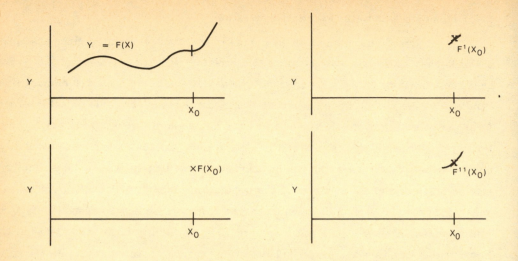

Figure 2.2.1. Sequence of figures indicate the contributions to the Taylor expansion of the functi $y = F(x)$ in (a); $F(x) = F(x_o) + (x - x_o)F'(x)|_{x=x_o} + 1/2(x - x_o)^2 F''(x)|_{x=x_o} + \cdots$ The first term $F(x_o)$ is indicated in (b); the linear term $(x - x_o) \cdot F'(x)|_{x=x_o}$ is shown (c) and the quadratic term in (d).

(2.2.1d). It is clear from this final graph that the approximation looks like the original function farther from the point $x = x_0$, than do the first two. Each successive approximation increases the region over which the approximate function and the actual function are essentially the same. It can be shown that for any well-behaved function the *infinite* series called the Taylor expansion gives an exact form for the function. This is in fact just the polynomial representation of the function centered on $x = x_0$. If $F(x)$ is sufficiently simple then it can be represented by its lowest order approximations.

The most familiar implementation of this expansion is to set $\alpha = F(x_0)$, $\beta = F'(x_0)$, and $F^{(n)}(x_0) = 0$ for all $n \geq 2$, to obtain the linear equation

$$y = \alpha + \beta x \tag{2.2.2}$$

where α is the intercept on the y axis and β is the slope. The values of α and β must be determined by experiment. We now assume that we can take a sequence of measurements (Y_j, X_j), $j = 1....,N$, in which the x-measurement is exact and error is concentrated in the y-measurement. With this assumption we use (2.2.2) to write

$$\xi_j = Y_j - (\alpha + \beta X_j) \tag{2.2.3}$$

for the deviation in the y-measurement from the anticipated (theoretical) expression. Graphically, ξ_j is the vertical distance from the point (Y_j, X_j) to the "optimum" straight-line representation of the data. The conditions for optimization are satisfied by selecting the constants α and β in a manner prescribed below.

A concrete example of a physical system for which these ideas are supremely useful is that of a simple electrical circuit. Let the set (Y,X) correspond to the measurement of the current through a resistor and the voltage across the resistor, respectively. If these two quantities are measured simultaneously then the temperature fluctuations in the material, estimation errors, etc., will cause variations in the repeated independent measurements of these two observables. A graph of current vs. voltage as one systematically varies the voltage across the resistor through a range of values gives a scatter of points on the graph. The regression of these data to the equation $y = \beta x$ is made yielding Ohm's law; one of the fundamental linear laws of physics. The parameter α is zero (experimentally) and β is interpreted as the average resistance of the resistor. The successes in the experimental determination of such linear laws in the physical sciences has motivated the wholesale adoption of such regression techniques throughout natural philosophy.

The method of least squares is used to obtain the best linear representation of the data, i.e., the optimum straightline through the data points. The method consists of defining the overall mean square deviation $\overline{\xi^2}$ for N measurements by

$$\overline{\xi^2} = \frac{1}{N} \sum_{j=1}^{N} \xi_j^2 \quad . \tag{2.2.4}$$

where (2.2.4) is a measure of the degree to which the data differ from the linear representation. This deviation is a function of the parameters α and β, i.e., $\overline{\xi^2} = \overline{\xi^2}(\alpha, \beta)$, which are to be chosen in such a way as to minimize the deviation from a linear function. This minimization condition corresponds to the mathematical constraints

$$\frac{\partial \overline{\xi^2}}{\partial \alpha} = 0 \quad \text{and} \quad \frac{\partial \overline{\xi^2}}{\partial \beta} = 0 \tag{2.2.5}$$

where the partial derivatives are taken with respect to α and β independently. The values of α and β obtained from these variational equations are

$$\alpha = \frac{\overline{X^2}\,\overline{Y} - \overline{X}\,\overline{YX}}{\overline{X^2} - \overline{X}^2} \tag{2.2.6}$$

$$\beta = \frac{\overline{YX} - \overline{Y}\,\overline{X}}{\overline{X^2} - \overline{X}^2} \tag{2.2.7}$$

which can be substituted into (2.2.2) to obtain the best fit to Y as a function of X. Here $\overline{X^2}$ is the mean square value of X and $\overline{XY} = 1/N \sum_{j=1}^{N} X_j Y_j$. Note that $\overline{XY} - \overline{X}\,\overline{Y}$ is the correlation between the pair-wise measurements of X and Y.

The minimization constraints (2.2.5) are obtained by varying $\overline{\xi^2}$ with respect to the parameters α and β and requiring that this variation vanish:

$$\delta \overline{\xi^2} = \frac{\partial \overline{\xi^2}}{\partial \alpha}\, \delta \alpha + \frac{\partial \overline{\xi^2}}{\partial \beta}\, \delta \beta = 0$$

Actually this only insures that the variation is an extremum not a minimum but that need not concern us here. Because α and β are independent parameters the condition that $\delta \overline{\xi^2}$ vanish can only be realized if the coefficients of the variation in $\alpha(\delta \alpha)$ and that in $\beta(\delta \beta)$ each vanish separately; thus yielding the expressions (2.2.5). To evaluate these coefficients we substitute (2.2.3) into the overall mean square deviation to obtain

$$\overline{\xi^2} = \frac{1}{N} \sum_{j=1}^{N} \left[Y_j - (\alpha + \beta X_j) \right]^2$$

$$= \overline{Y^2} + \alpha^2 + \beta^2\,\overline{X^2} - 2\alpha \overline{Y} + 2\alpha\beta \overline{X} - 2\beta \overline{YX} \quad .$$

Then the appropriate derivatives are

$$\frac{\partial \overline{\xi^2}}{\partial \alpha} = 2\left[\alpha - \overline{Y} + \beta \overline{X} \right] = 0$$

$$\frac{\partial \overline{\xi^2}}{\partial \beta} = 2[\beta \overline{X^2} + \alpha X - \overline{YX}] = 0$$

which gives the two algebraic equations

$$Y = \alpha + \beta X$$

$$\overline{YX} = \alpha X + \beta \overline{X^2} \ .$$

We now have two equations in two unknown parameters α and β which can be solved in terms of the measured quantities X, Y, $\overline{XY}$ and $\overline{X^2}$ to yield (2.2.6) and (2.2.7).

The statistical method analogous to the least square procedure described above is the maximum likelihood of a variate having a *linear regression* about a single observable. Regression is actually a procedure by which to make a least biased prediction based on past experimental work. In the following discussion we shall be concerned with a variate Y having a normal distribution in which the mean is a function of the observables, and the variance is assumed to be independent of the mean. The mean $F(x)$, where "X" is the set of observables, is called the *regression function*; this function represents a curve if x consists of a single variable, a surface if x consists of two variables, and a hyper-surface for an x consisting of more than two variables [see e.g., Mood, 1950].

As in the previous discussion we assume that the regression function $F(x)$ is linear in a single observable with probability density for y given by

$$P(y; \alpha, \beta, \sigma^2, x) = \frac{1}{\sqrt{2\pi}\,\sigma} \exp\left\{-\frac{1}{2\sigma^2}[y-(\alpha+\beta x)]^2\right\} \ . \tag{2.2.8}$$

We have indicated the dependence of the probability density on α, β, σ^2 and x explicitly. Equation (2.2.8) represents a one-parameter family of normal distributions since for each value of x, y is normally distributed with average value $\alpha + \beta x$ and variance σ^2. We use the normal distribution itself to estimate the values of α, β and σ^2 using N sample points $j=1, \ldots, N$ for (Y_j, X_j). The method of maximum likelihood is used to estimate these parameters. The likelihood is [note the similarity to (2.1.7)]

$$L = \left(\frac{1}{2\pi\sigma^2}\right)^{\frac{N}{2}} \prod_{j=1}^{N} \exp\left\{-\frac{1}{2\sigma^2}\left[Y_j-(\alpha+\beta X_j)\right]^2\right\} \tag{2.2.9}$$

and following the procedure used earlier we take the logarithm of L:

$$\log L = -\frac{N}{2}\log(2\pi\sigma^2) - \frac{1}{2\sigma^2}\sum_{j=1}^{N}\left[Y_j-(\alpha+\beta X_j)\right]^2 \ . \tag{2.2.10}$$

As in the method of least squares we vary $\log L$ with respect to α and β and set the result equal to zero, since for L to be a maximum the term in the exponent must be a minimum. Thus we solve the resulting linear equations to obtain results having the same form as (2.2.6) and (2.2.7). We denote these solutions by $\hat{\alpha}$, $\hat{\beta}$ and $\hat{\sigma}^2$ to indicate that these are the point estimators of the unknown parameters:

$$\hat{\alpha} = \frac{\overline{Y}\,\overline{X^2} - \overline{YX}\,\overline{X}}{\overline{X^2} - \overline{X}^2} = \overline{Y} - \hat{\beta}\overline{X} \tag{2.2.11a}$$

$$\hat{\beta} = \frac{\overline{YX} - \overline{Y}\,\overline{X}}{\overline{X^2} - \overline{X}^2} \tag{2.2.11b}$$

$$\hat{\sigma}^2 = \frac{1}{N} \sum_{j=1}^{N} \left[Y_j - (\hat{\alpha} + \hat{\beta}X_j) \right]^2 \tag{2.2.11c}$$

The Y_j have a normal distribution of values and the estimators are linear functions of Y_j so it follows that $\hat{\alpha}$ and $\hat{\beta}$ themselves have a bivariate normal distribution of values. This distribution can be specified by finding the averages, variances and covariances of $\hat{\alpha}$, $\hat{\beta}$, i.e., for the variates $(\hat{\alpha} - \alpha)/\sigma_\alpha$ and $(\hat{\beta} - \beta)/\sigma_\beta$.

It is difficult to understand the implications of these ideas without directly working with experimental data sets. Since it is not possible to provide raw data in a book without presenting large tables of numbers, (which are never read), we have opted for providing the reader with a computer code that will generate a DATA SET. As part of this code a linear regression to the generated data is given. Thus the reader inputs the parameters α and β and the number of experimental observations N. The code then generates a set of Gauss random variables $\{Y_j, X_j\}, j = 1, \cdots, N$ related by (2.2.3) and plots them on an (y, x) graph. The theoretical estimates $\hat{\alpha}$ and $\hat{\beta}$ are then calculated from the data and the optimum linear fit to the data graphed. A comparison of the input intercept (α) and slope (β) with the linear regression values $\hat{\alpha}$ and $\hat{\beta}$ as one changes the number of experimental observations N should be instructive.

Following Mood (1950) let us assume that a linear regression function $y = \alpha + \beta x$ has been estimated by $y = \hat{\alpha} + \hat{\beta}x$ using N observations. We now want to *predict* the value of y given some particular measurement $x = x_0$. Thus if Y is the adult height of the sons in a population and X is the corresponding height of the fathers, a sequence of observations will provide the data for forming estimates of $\hat{\alpha}$ and $\hat{\beta}$ for a linear regression function. An anxious father of height x_0 wishes to predict his son's most probable height as an adult. The predicted height he learns is $y_0 = \hat{\alpha} + \hat{\beta}x_0$ using the sample data. This prediction is without value unless we can also associate a *prediction interval* in analogy with a *confidence interval*. This can be done using a t-distribution with $N-2$ degrees of freedom and which involves no unkown parameters. We do not describe this technique here since this would lead us further into the realm of traditional statistical analysis and divert us from our main purpose, see e.g., Bevington (1969).

The reverse of the above problem of prediction is the associated problem of *discrimination* which is an estimation problem. For prediction one selects a desired value of x, say x_0 and wishes to predict the associated value of y on the basis of the estimates of α, β and σ. In discrimination one wishes to estimate x_0 after having observed Y. As discussed by Mood (1950) the general class of biological assay problems are of this latter character as are general classification problems. As an example he cites the measurement of skulls Y of known ages X, made by anthropologists, to then estimate the age x_0 of a skull of unknown age with the set of measurements Y'. Also taxonomists use the technique to discriminate between varieties of plants with quite similar appearances. Again employing a maximum likelihood argument one obtains $x_0 = \hat{\beta}^{-1}(Y' - \hat{\alpha})$, where the measurements are used to estimate $\hat{\alpha}$ and $\hat{\beta}$. Again this estimation is without value unless we can associate a *confidence interval* with x_0 and again this can be accomplished with a t-distribtuion.

In summary we reiterate that in practice there appears to be two concepts which characterize the state of the art in the application of statistics to natural philosophy. The first is the Gauss distribution and the second is the method of least squares, both of which are based on linear concepts in that they require the minimization of a quadratic quantity. It should also be emphasized that the arguments presented in this section are not limited to the linear form of $F(x)$ obtained from the Taylor expansion (2.2.1). One may elect not to expand $F(x)$, which in general is a nonlinear function of its parameters. In that case the procedure that is employed is a method of *nonlinear least squares*; see eg. Bevington (1969). Again one minimizes the deviation from the theoretical $F(x)$ of a sequence of N measurements (Y_j, X_j)

$$\overline{\xi^2} = \frac{1}{N} \sum_{j=1}^{N} \left[Y_j - F(X_j) \right]^2 \tag{2.2.12}$$

with respect to a set of parameters $\{\alpha_k\}$, $k=1, 2, \cdots, M$:

$$\frac{\partial}{\partial \alpha_k} \overline{\xi^2} = \frac{\partial}{\partial \alpha_k} \frac{1}{N} \sum_{j=1}^{N} [Y_j - F(X_j)]^2 = 0 \ . \tag{2.2.13}$$

Here $\overline{\xi^2}$ must be considered a continuous function of the M parameters α_k describing a hypersurface in an M-dimensional space, and the space must be searched for the appropriate minimum value of $\overline{\xi^2}$. However the criteria used, (2.2.13) is a direct application of the linear concepts developed earlier. It is still a method of least squares and as such is a linear concept. For example (2.2.13) will result in a system of M linear equations for the moments of X with coefficients given by the α_k's. The solution to this linear system enables us to express the values of the optimum set of parameters in terms of the moments of Y and X using the data set. Linearity expresses the relationship between the unknown parameters and the moments of the observables.

2.2.2 Time Series

We have until now restricted our discussion to discrete data sets obtained from a sequence of independent measurements. The form in which the data is obtained is often dictated by the measuring instruments that are available rather than the variable of interest. On the other hand the variable of interest often cannot be directly measured and must be deduced from those measurements which are possible. For example, a physician does not have a meter to tell him the state of a patient's health, so his judgment as to that state is based on indirect evidence. This evidence consists of a collection of pieces of information that may contain electrocardiograms, blood pressure readings, laboratory values for liver enzymes, blood sugar, kidney function and on and on. Each of these measurements is individually compared with an experimental standard that the medical community has developed over the past few thousand years (most standards are of a more recent origin). If all the indicators fall within pre-established limits (sometimes these limits are not as stable from physician to physician as one might like) then the physician concludes that the patient is free of disease. The implication is that if the patient is *disease-free* then that person is healthy. This particular point of view is presently being called into question by parts of the medical community, see e.g., Weiner (1984) for a scholarly historical discussion from the point of view of psychosomatic medicine and Dossey (1982) for a popularized presentation of new thoughts on the definition of health, based on the modern-physics paradigm. In any event, by using the disease-free definition of health, and this is what most practicing physicians do, one is forced into an operational representation of health that is a composite. At a given point in time, or over a relatively short time interval, the set of medical parameters *defines* that state of health of the individual. The fancy statistical procedures we have so far discussed do not directly aid the physician in the formation of a diagnosis since the composite picture constructed is much too sparse to be meaningful from a data sampling perspective. However, if one of these data points exceeds the phenomenologically established limits, then either subsidiary tests, to decide between "causal" alternatives, or more intensive tests, to accumulate the necessary statistical data base on which to formulate a recommendation, are ordered.

Once a person has been tentatively diagnosed as severly ill, say with a life-threatening arrhythmia, the above point measurements may no longer be an adequate format for gaining information. Information on the primary indicators of the state of the cardiovascular system must be continuous in order that a transition into a life threatening situation can be anticipated and hopefully averted. This continual monitoring provides information in the form of a time series, say for example the electrocardiogram (ECG) trace in a critical care unit which records the continuous operation of the heart and its support systems on a real time basis. The question is: *How is this new flood of information utilized?*

The information contained in an ECG record is used in a correlational way, that is to say, different time trace patterns are associated with different conditions of the heart. For example a normal ventricular pattern consists of a QRS pulse complex followed after a certain time interval by a normal T wave. The atrial ECG is represented by a P wave that triggers the ventrical and the total ECG by the composite of the P-QRS-T cycle taken as a unit [Goldberger and Goldberger (1981)]. Any disruption of this pattern is interpreted as (correlated with) a physical mechanism or process in the conduction path that is not normally present. As pointed out by Goldberger and Goldberger (1981), the ECG is a recording of cardiac electrical activity and does not in any way measure the mechanical function of the heart, i.e., how well the heart is operating as a pump. Thus a patient in acute pulmonary edema may have a normal ECG. On the other hand it is often possible to *infer* a specific structural diagnosis such as mitral stenosis, pulmonary embolism, as myocardial infraction from the ECG because *typical electrical abnormalities* may develop in such patients. Therefore the use of the information contained in an ECG has to do with pattern recognition. The physician has been trained to interpret (associate) certain patterns in the ECG trace with certain pathologies and in this regard his experience provides a matched filtering between the continuous flow of information and a diagnosis. The physician acts as a data analyzer, as do we all, at least some of the time.

Let us denote the ECG time trace by the function $A(t)$ which is the amplitude of the deflection of the recording needle as a function of time. This deflection is assumed to be proportional to the electrical activity traveling across the surface of the heart. As we pointed out above, there is an intrinsically oscillatory component to the ECG record $A(t)$, but many other biological and behavioral phenomena show a similar intrinsic oscillatory behavior. Thus we wish to develop a data processing technique which will enable the investigator to extract the information required with a minimum of effort -- at least at the time the information is needed. Visual inspection alone of the time histories of the phenomenon, such as employed by our above cardiologist, does not provide sufficient information on noise content, real periodicities or the distribution of periodicities manifest in a system. A quantity called the autocorrelation function (also known as the covariance function or just covariance) and the power spectral density (PSD) provide useful ways to encapsulate vast amounts of data into a readily assimilated format.

The autocorrelation function provides a way to use the data obtained at one time to determine the influence of the process on itself at a later time. It is an average measure of the relation of the value of a random process at one instant of time, $A(t)$ say, to the value at another instant of time τ seconds later, $A(t+\tau)$. If we have a data record extending continuously over the time interval $(0, T)$, then the autocorrelation function is defined as

$$C_{AA}(\tau) \equiv \lim_{T \to \infty} \frac{1}{T} \int_0^T A(t) A(t+\tau) dt \quad . \tag{2.2.14}$$

Note that for a finite sample length, i.e., for T finite, the integral defines an *estimate* for the autocorrelation function so that $C_{AA}(\tau) = \lim_{T \to \infty} C_{AA}(\tau, T)$. In Figure (2.2.2) a sample time history of $A(t)$ is given along with the displaced time trace $A(t+\tau)$. (Note that $A(t)$ in this figure is *not* intended to represent an ECG.) The point by point product of these two traces appears in (2.2.14), i.e., the product of the "signal" and the signal delayed (displaced in time) by an interval τ. The product is then averaged; integrated over the interval $(0,T)$ and divided by T. It is often assumed in theoretical studies that the time averages of the data for a sufficiently large T can be replaced by an average over an ensemble distribution function $P(a;t \mid a_0)$. This is the so-called ergodic hypothesis. The probability that the dynamic variable $A(t)$ has a value in the interval $(a, a+da)$ at time t given an initial value $a_0 = A(0)$ is $P(a;t \mid a_0) da$. We have already discussed replacing an average over a data set by an ensemble average. However the ensemble distribution given here is somewhat different in that it depends on the initial state of the system under study. These averages are therfore called conditional, or two-point averages and are necessary in order to calculate an autocorrelation function.

A "typical" autocorrelation function is depicted in Figure (2.2.3) where the time lag τ is varied over a range of values and the signal is generated by a random process known to have a short memory time τ_c. A monochromatic sine wave, or any other

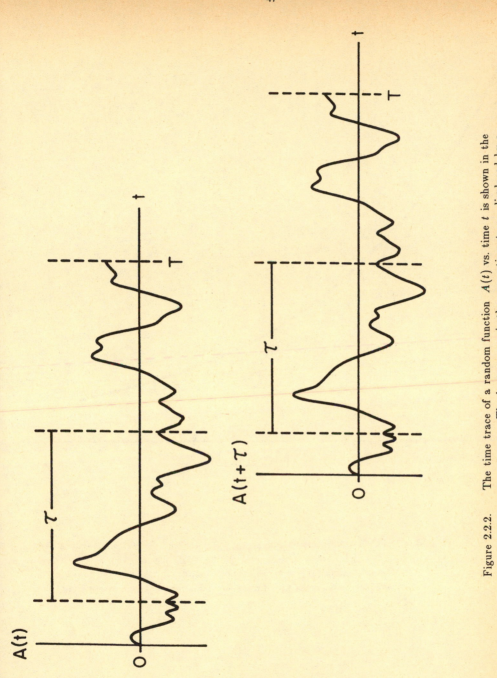

Figure 2.2.2.   The time trace of a random function $A(t)$ vs. time t is shown in the upper curve.   The lower curve is the same time trace displaced by a time interval τ.   The product of these two functions when averaged yield an estimate of the autocorrelation function $C_{AA}(\tau,T)$.

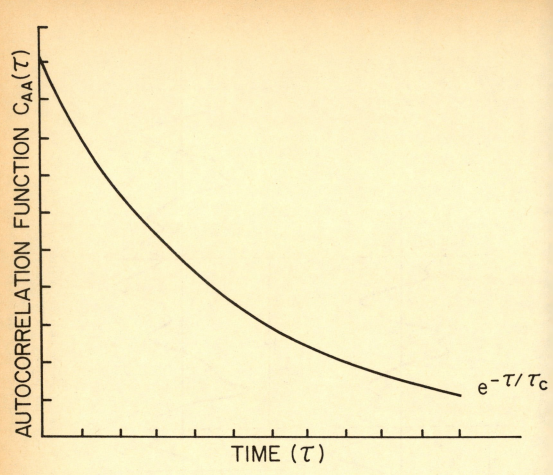

Figure 2.2.3.   The autocorrelation function $C_{AA}(\tau)$ for the typical time traces depicted in Figure 2.2.2 assuming the fluctuations are exponentially correlated in time $[\exp(-\tau/\tau_c)]$. The constant τ_c is the time required for
$C_{AA}(\tau)$ to decrease by a factor $1/e$, this is the decorrelation time.

determinisitc data, would have an autocorrelation function which persists over all time displacements. Thus the autocorrelation function can provide an indication of deterministic data embedded in a random background.

Similar comments also apply when the data set is discrete rather than continuous. In this case we denote the interval between samples as $\Delta(=T/N)$ for N samples and r as the lag or delay number so that the estimated autocorrelation function is

$$C_{AA}(r\Delta,N) = \frac{1}{N-r} \sum_{j=1}^{N-r} A_j A_{j+r} \quad ,r = 0, 1, ..., m \tag{2.2.15}$$

and m is the maximum lag number. Again the autocorrelation function is defined as the limit of (2.2.15) when $N \to \infty$. These considerations have been discussed by Wiener (1949) at great length and his book on time series is still recommended as a text from which to capture a master's style of a certain way of mathematically viewing such problems.

In our discussion of the random walk problem in §2.1 we introduced the definition of a Fourier transform [cf. (2.1.17)] over a finite interval, i.e., the variation in phase over the interval $[-\pi, \pi]$. The idea we used was that the Fourier transform of $\cos^n \phi$ over the angle ϕ in (2.1.18) extracted out (filtered) those values of n for which the integral is non-zero. This same idea can be applied to the autocorrelation function since it provides a measure of the periodicities (frequencies) in the time series. This frequency content can be explicitly displayed by taking the Fourier transform of $C_{AA}(\tau)$:

$$S_{AA}(\omega) = \frac{1}{\pi} \int_{-\infty}^{\infty} C_{AA}(\tau)e^{-i\omega\tau} d\tau$$

$$= \frac{2}{\pi} \int_{0}^{\infty} C_{AA}(\tau)\cos\omega\tau\, d\tau \tag{2.2.16}$$

where $S_{AA}(\omega)$ is the power spectral density (PSD) of the time series $A(t)$. Equation (2.2.16) relates the autocorrelations function to the power spectral density and is known as the *Wiener-Khinchine* relation. One example of the use of this relation is provided by Figure (2.2.4) where the exponential form of the autocorrelation function $C_{AA}(\tau) = e^{-\tau/\tau_c}$ used in Figure (2.2.3) yields a frequency spectrum of the Cauchy form

$$S_{AA}(\omega) = \frac{1}{\pi} \frac{\tau_c}{1 + \omega^2 \tau_c^2} \quad . \tag{2.2.17}$$

At high frequencies the spectrum is seen to fall-off as ω^{-2}. Basar (1976), among

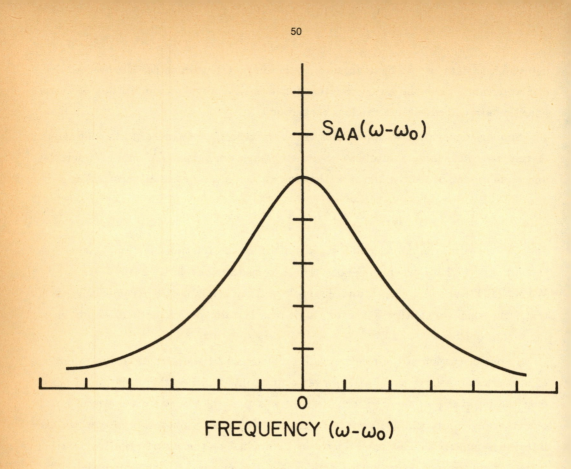

Figure 2.2.4.   The power spectral density $S_{AA}(\omega)$ is graphed as a function of frequency for the exponential correlation function depicted in Figure 2.2.3. The spectrum for such an autocorrelation function is the Cauchy distribution (also called the Lorentz distribution).

others, has applied these techniques to the analysis of many biological phenomena including electrical signals from the brain.

Again there is a discrete counterpart to the PSD that is based on taking a discrete Fourier transform of the discrete autocorrelation function. There are a number of mathematical subtleties involved in the manipulation of such series and in the replacement of such series by integrals such as (2.2.16). We will not concern ourselves here with these details since they are available in any of a number of excellent text books, see e.g., the classic work of Zygmund (1935). However it is worthwhile to explore the frequency content of continuoous time series so as to develop an insight into the possible dynamic behavior associated with a spectrum. For this purpose we suggest that the reader exploit the computer program TIME SERIES provided in the Appendix.

In the TIME SERIES computer program a continuous time trace is generated for the function

$$A(t) = \sum_{k=1}^{N} A_k \cos\omega_k \tau \tag{2.2.18}$$

where the set $\{A_k\}$ is generated by a Gauss distribution of random numbers. The user is able to independently specify the relation between the frequency ω_k and the mode number k, i.e., the so-called dispersion relation, as well as the number of modes N. Thus (2.2.18) is a Gauss random process in time. An example of how this time trace is used is to construct the autocorrelation function

$$C_{AA}(\tau) = \; < A(t)\, A(t+\tau)> \tag{2.2.19}$$

where the brackets denote an average over an ensemble of realization of the set $\{A_k\}$. Thus for a stationary process the autocorrelation function is

$$C_{AA}(\tau) = \sum_{k=1}^{N} < A_k^2 > \cos\omega_k \tau \tag{2.2.20}$$

so that comparing (2.2.20) with (2.2.16) we have the spectrum

$$< A_k^2 > = \frac{1}{2}\frac{\tau_c}{1+\omega_k^2 \tau_c^2} \; . \tag{2.2.21}$$

and

$$C_{AA}(\tau) = \frac{\tau_c}{2}\sum_{k=1}^{N}\frac{\cos\omega_k \tau}{1+\omega_k^2 \tau_c^2} \tag{2.2.22}$$

is the correlation function corresponding to the exponential relaxation in Figure (2.2.3) in the limit $N \rightarrow \infty$. Precisely how the spectrum (2.2.21) and the correlation function (2.2.22) vary with the number of modes N in the spectral representation of $A(t)$ is quite illuminating. The TIME SERIES code allows the user to explore other choices of the set $\{A_k\}$ to gain familiarity with the visual appearance of $A(t)$ as its spectral content is varied.

2.3 Langevin Equations

In the previous sections we focused our attention on the distribution of errors obtained in a sequence of measurements of a given quantity. Is it in fact necessary that one contend with a statistical error in a measurement? Is it not possible to increase the resolution of one's instrument sufficiently that the observed random error vanishes? Here again, we are not considering quantum mechanical effects, but are concerned with macroscopic phenomena. If one assumes that the variable of interest can be isolated from the environment and the resolution of the measuring instrument can be made arbitrarily fine then the error distribution can become a delta function centered on zero. If however, the quantity being measured cannot be isolated from the environment then even with the finest resolution, a non-systematic variation from measurement to measurement will be observed due to the coupling of the observable to the unobserved background.

In some of our earlier discussions we emphasized the desire on the part of the scientist to predict the future. The success of the predictions then become the measure of the level of understanding of the underlying process. If an economic forecaster predicts an upturn in the stock market, and it occurs, we gain confidence in the economist's understanding of the market and his ability to make predictions. On an individual basis we may have a great deal of success with our investments, but this is distinct from a formal understanding of the mechanisms at work in the market place. Some of the operations research studies that have been done indicate that, over the long term, a random walk model of the stock market does no worse than other models which purport to be "explanatory." In his book, Cootner (1964) compiles a number of important papers on *The Random Character of Stock Market Prices* starting with certain random walk models. The reason an individual broker may be successful in this random environment stems from the broker's belief that there are patterns embedded in this random process, and by "technical analysis" he can uncover when to "buy low and sell high" and pass this advice along to his customers.

The above point of view asserts that in the main, the motion of the market is a random process driven by external events (coupling to the environment). The events are international and national issues related to war and peace, economic successes and failures of smaller segments of the economy, crop failures, shifts in the political perspective, interest rates, inflation, population trends and more. All these activities (causes) tend to mask any underlying market mechanisms that might have the character of a "social or economic law." Thus, as pointed out by Montroll and Badger (1974), the response of the buyers and sellers of securities varies from time to time in such a way that the market seems to be responding to a random force and behaves in a manner not unlike a molecule in a fluid being buffeted about by many other lighter molecules. This process, known as Brownian motion, is discussed in the next section. One can in fact have an underlying force (theoretically) which guides the short term apparently random activity of the market and which can be extracted using the proper non-stationary data processing techniques.

This example summarizes the way in which the concepts of randomness, fluctuations, stochastics and so on have come to be understood in natural philosophy. Let us then summarize in a fairly formal way the content of this understanding [see e.g., Lindenberg and West (1984) and West and Lindenberg (1983)]. We assume that the state of the dynamic system can be described by N variables $\{X_\alpha(t)\}$, $\alpha = 1, 2,..., N$ where, for example the state vector $\mathbf{X}(t)$ might represent the prices of a selected set of stocks on the NYSE along with all the factors that have been identified with being able to influence their market value. For a sufficiently complex market the limit $N \to \infty$ can be taken. We refer to the evolution of $\mathbf{X}(t)$ away from its initial state $\mathbf{X}(0)$ as a general flow field. The evolution of the flow field can be described by a deterministic trajectory $\Gamma_t(\hat{\mathbf{x}})$ in an N-dimensional phase space with the set of axes $\hat{\mathbf{x}} = \{\hat{x}_\alpha\}$ corresponding to the full set $\{X_\alpha(t)\}$ of dynamic variables. The curve $\Gamma_t(\hat{\mathbf{x}})$ is indexed by the time t which is a parameter in the phase space. A trajectory begins at a specified point $\hat{\mathbf{x}}_0 = \{X_\alpha(0)\}$ and describes the evolution of the flow towards its final state $\{X_\alpha(\infty)\}$. In practical calculations only a relatively small number of variables $(M \ll N)$ are used to represent the flow field. For the stock market this means that although all these other factors in actual fact are part of the operation of the market, only the prices of the stocks (or logarithm of the price) are used to characterize the condition of the market at a given point in time.[5] Mathematically this means that one is interested only in the projected trajectory

$$\Gamma_t(\mathbf{x}) = P\,\Gamma_t(\hat{\mathbf{x}})$$

where $\mathbf{x} = \{x_\alpha\}$ with $\alpha = 1, 2,..., M$ and P denotes an appropriate integration over the variables $x_{M+1}, x_{M+2}, ... x_N$. The operation denoted by this equation is a projection from an N-dimensional space denoting the full system down to an M-dimensional space denoting only the variables of immediate interest.

If we consider two trajectories with identical initial values $X_1(0)$, $X_2(0)$, ..., $X_M(0)$ for the observed degrees of freedom but with different initial values $X_{M+1}(0)$, ..., $X_N(0)$ for the unobserved degrees of freedom, the two projected trajectories may be different, since they each interact with the *unobserved* but *different* initial states that have been projected out. Thus two trajectories ostensibly initiated from the same state $X_1(0)$, ..., $X_M(0)$ in the reduced phase space can follow different orbits in the reduced space. We refer to the instantaneous differences between two such trajectories as fluctuations and presume that they admit of a statistical description. The statistics enter through the specification of the initial conditions of the eliminated degrees of freedom: since these are not observable one can specify a distribution of initial conditions $X_{M+1}(0)$, ..., $X_N(0)$ consistent with the "macroscopic" initial state of the flow $\mathbf{X}_0$. This is the distinction between macroscopic and microscopic motion made by the classical physicist and chemist; macroscopic is what is observed by experiment, microscopic is not directly observed but has manifest influences on the observational level.

This style of thought always leads one to a set of dynamic equations for the physical observables which is *stochastic*. These stochastic differential equations can be linear as we describe in this section or they can be nonlinear as we discuss subsequently. The point is that in order to reduce our description from a large number of degrees of freedom down to one of manageable size we have had to introduce fluctuations. The apparent price of this mathematical tractability is the loss of *predictability*. Can we in fact avoid paying this price? Can we not just truncate the deterministic equations of motion at some level and merely ignore the coupling to the other degrees of freedom? Isn't the resulting deterministic description to be preferred over the more exact stochastic description? The short answer is no. The long answer, given later, is that in a nonlinear system even a few degrees of freedom can lead to chaotic or "random" motion and fluctuations are therefore unavoidable. Of course the stochastic motion resulting from the "coarse-graining" described above and that resulting directly from the nonlinear character of the system which is described in subsequent sections, may or may not have anything to do with one another as we will discover in the sequel.

Let us consider a complex event that in principle can be characterized by (N+1) variables denoted by $\mathbf{X} = (X_0, X_1, \cdots, X_N)$. We assume that the dynamic variables satisfy the linear equations of motion

$$\frac{d\mathbf{X}}{dt} = \mathbf{A}\,\mathbf{X}$$

(2.3.1)

where $\mathbf{A}$ is an $(N+1)\times(N+1)$ matrix whose constant elements determine the coupling among the dynamic variables. Let us further assume that the measurement technique only addresses the single variable $X_0(t)$, say the profit earned by a single stock in the market. Therefore we refer to $X_0(t)$ as the observable and the remaining variables ($X_1,..., X_N) \equiv \mathbf{Y}(t)$ as unobservables. As in our example all the factors influencing the profitability of a given stock are not recorded with the price of the stock. Therefore the individual speculator, although he may know a few of the factors modifying the stock price, is unaware of the majority of them, i.e., they go unobserved. The equation of motion (2.3.1) can be partitioned as follows to emphasize this separation,

$$\frac{d}{dt}\begin{pmatrix}X_0 \\ \mathbf{Y}\end{pmatrix} = \begin{bmatrix} a & \mathbf{b} \\ \mathbf{c} & \mathbf{d} \end{bmatrix}\begin{bmatrix}X_0 \\ \mathbf{Y}\end{bmatrix} \tag{2.3.2}$$

where a is a constant, $\mathbf{b} = (b_1,...,b_N)$ is a vector with constant elements, as are the vector $\mathbf{c}^+ = (c_1,...,c_N)$ and the $N \times N$ matrix $\mathbf{d}$.

The linear nature of the equations (2.3.2) enables us to directly integrate those for the unobservables:

$$\frac{d\mathbf{Y}}{dt} = \mathbf{c}\, X_0 + \mathbf{d}\,\mathbf{Y} \tag{2.3.3}$$

to obtain

$$\mathbf{Y}(t) = e^{\mathbf{d}t}\,\mathbf{Y}(0) + \int_0^t e^{\mathbf{d}(t-\tau)}\,\mathbf{c}\,X_0(\tau)d\tau \tag{2.3.4}$$

where $\mathbf{Y}(0)$ is the initial value of the vector of unobservables $\mathbf{Y}(t)$.[6] This solution (2.3.4) can be substituted into the equation of motion for $X_0(t)$ to obtain

$$\frac{dX_0(t)}{dt} = aX_0(t) + f(t) + \int_0^t K(t-\tau)X_0(\tau)d\tau \tag{2.3.5}$$

where

$$f(t) \equiv \mathbf{b}\, e^{\mathbf{d}t}\,\mathbf{Y}(0) \tag{2.3.6}$$

and

$$K(t-\tau) \equiv \mathbf{b}\, e^{\mathbf{d}(t-\tau)}\,\mathbf{c} \tag{2.3.7}$$

Equation (2.3.5) is no longer explicitly dependent on the unobservables $\mathbf{Y}(t)$. The vestige of this dependence is through the initial conditions $\mathbf{Y}(0)$ and the elements of the coupling matrices $\mathbf{b},\mathbf{c}$ and $\mathbf{d}$. We see that the coupling to the environment has produced a memory in the evolution of the observable X_0, that is to say, the profit of a stock at time t is determined by past events as well as present ones. The integral term is the manifestation of the back-reaction of the environment onto the observable at time t due to an interaction at the earlier times $t - \tau$, i.e., economic forces are responsive to the present state of the market and influence its future development by accommodating this state. Such time retarded effects are not uncommon in other models of biological and behavioral phenomena.

The remaining term in (2.3.5) due to the environment is $f(t)$. This time-dependent function depends on the initial state of the unobservables. One is tempted to specify the initial state as $Y(0) = 0$ so that this term no longer contributes to the evolution of $X_0(t)$. However this specification is inconsistent with the assumption that $Y(t)$ is unobservable and it is clear that such a specification would not correspond to any realizable state of the market. In fact since only $X_0(t)$ is being observed, only the initial state $X_0(0)$ can be specified by the experiment. Stated differently, if we specify the initial state of the observable $X_0(0)$ then, as discussed previously, there is a collection of initial states of the unobservables that are consistent with $X_0(0)$. Thus there is a fundamental *uncertainty* in the specification of $Y(0)$ which may be characterized by associating a distribution function with the possible initial states of the unobserved degrees of freedom, i.e., the initial state of the market. Rewriting (2.3.6) in the schematic form

$$f(t) = \sum_{j=1}^{N} \alpha_j(t) Y_j(0) \tag{2.3.8}$$

if: (1) the $Y_j(0)$ have the same distribution for all j, (2) the $\alpha_j(t)$ are harmonic functions of time and (3) N becomes very large, then $f(t)$ is a random function having Gauss statistics. Even when these conditions are not met $f(t)$ represents a fluctuating function since the $Y_j(0)$'s are random; therefore (2.3.5) is a stochastic differential equation. In the physical sciences (2.3.5) would be called a "generalized Langevin equation" after P. Langevin (1908) who first introduced such equations into physics.

There can be conditions on the properties of the coupling terms, i.e., the elements of **d**, **b** and **c**, under which the memory kernel in (2.3.5) reduces to a delta-function:

$$K(t - \tau) = 2\lambda \, \delta(t - \tau) \tag{2.3.9}$$

where λ is a constant. In physical systems $K(t - \tau)$ would not be a delta-function, but rather a function having a non-zero value over a time interval τ_c very much shorter than the characteristic time scale of the system, i.e., $\tau_c \ll \dfrac{1}{|a|}$. In this situation the equation of motion for the observable (2.3.5) becomes

$$\frac{dX_0}{dt} + (\lambda - a) X_0 = f(t) \quad . \tag{2.3.10}$$

The solution to (2.3.10) is then given by

$$X_0(t) = e^{-(\lambda - a)t} X_0(0) + \int_0^t e^{-(\lambda - a)(t - \tau)} f(\tau) d\tau \tag{2.3.11}$$

where the integral term is a fluctuating quantity in time.

Let us for now assume that $f(t)$ is a zero-centered Gauss process, delta-correlated in time. Indicating an average over an ensemble of realizations of the fluctuations $f(t)$ by a bracket $<-->$ we denote these assumptions by

$$<f(t)> = 0 \tag{2.3.12a}$$

$$<f(t_1) f(t_2)> = 2D\delta(t_1 - t_2) \tag{2.3.12b}$$

where D is the spectral strength of the fluctuations. The average profitability of our stock $X_0(t)$ is then given by

$$<X_0(t)> = e^{-(\lambda - a)t} X_0(0) \tag{2.3.13}$$

where we have used (2.3.12a) to eliminate the integral term in (2.3.11). If $\lambda > a > 0$, indicating that the market tends to extract profit (λ) more strongly than the natural predisposition for the profit to grow (a), then the asymptotic value of $<X_0(t)>$ vanishes:

$$\lim_{t \to \infty} <X_0(t)> = 0 \tag{2.3.14}$$

since $e^{-(\lambda - a)t} \to 0$ as $t \to \infty$. Thus there is no bias in the profitability as the system achieves equilibrium since the average profit is zero.

We can also calculate the mean square value of the profit by multiplying the solution (2.3.11) by itself and averaging over the same ensemble as above:

$$<X_0^2(t)> = e^{-2(\lambda - a)t} X_0^2(0) + \frac{D}{\lambda - a} [1 - e^{-2(\lambda - a)t}] \quad . \tag{2.3.15}$$

In the long time limit the mean square profit (2.3.15) reduces to

$$\lim_{t \to \infty} <X_0^2(t)> = \frac{D}{\lambda - a} \tag{2.3.16}$$

so that even though the averge value of the profit vanishes asymptotically there remains a finite mean square value which is directly proportional to the spectral strength of the induced flucutations and inversely proportional to the strength of the dissipation. One can interpret (2.3.16) as a balanace between the influx of profit due to fluctuating market conditions and the efflux of profit due to average market operation:

$$\lim_{t \to \infty} <X_0^2(t)> = \frac{influx}{efflux} \quad . \tag{2.3.17}$$

The relative sizes of these terms determine the relative certainty with which one knows the equilibrium value of the profit $X_0(t)$.

Because the solution (2.3.11) is linear in the fluctuations the statistics of $X_0(t)$ are the same as those of $f(t)$. Thus we expect that the statistics of $X_0(t)$ are described by a zero-centered Gauss distribution with a variance given by (2.3.16):

$$P_{ss}(x) = \left(\frac{\lambda - a}{2\pi D} \right)^{1/2} \exp\left\{ -\frac{\lambda - a}{2D} x^2 \right\} \tag{2.3.18}$$

where $P_{ss}(x)dx$ is the probability that $X_0(t)$ has a value in the interval $(x, x + dx)$ in the equilibrium state. We will in fact prove this in a subsequent section. Variations of this simple model to include the effect of "inflation" and "insiders" in the market are also interesting to examine, see eg. West (1974) and (1980).

If there are variations in the observations due to random measurement error, then these variations will superpose on the integral term in (2.3.11). Although the measurement errors have a Gauss distribution, the fluctuations induced by the coupling to the environment *may not*, so that the distribution of the net fluctuations cannot be specified without more detailed knowledge of the system and its environment. In particular it may well *not* have the Gauss form, although in the illustrative example used above we did assume $f(t)$ to have a Gauss distribution, and since the system is linear the system response also had such statistics. If the dynamics are specified by a non-linear equation of motion having a fluctuating driving force $f(t)$, then even if $f(t)$ has Gauss statistics the system response will not.

2.4 Chain Conditions and Fokker-Planck Equations

The physical scientist's literature on stochastic processes has its modern origin in the theories of Einstein (1905) and Smoluchowski (1906, 1916) concerning Brownian motion. The process is named in memory of the observations of the English botanist Robert Brown (1829), who noted the erratic motion of pollen grains suspended in fluids. The actual effect was first observed by the Dutch physician, Jan Ingen-Housz (1784) who, while in the Austrian court of Empress Maria Theresa, observed that finely powered charcoal floating on an alcohol surface executed a highly erratic random motion. His (Ingen-Housz) other accomplishments include the vaccinating of the family of George III of England against smallpox, and the observation that sunlight is necessary for the process by which plants grow and purify an atmosphere contaminated by the breathing of animals, i.e., photosynthesis. Einsteins' theory of the random motion of "large" particles in fluids, Brownian motion, was used as a proof of the existence of molecules, and Perrin's (1910) and Svedberg's (1912) quantitative measurements of particle displacements as a function of time yielded a numerical value for Avogadro's number, i.e., the number of molecules in a gram-molecule of air.

The chain equation

$$P(y_2; t \,|y_1) = \int P(y_2; t \,|y;\tau) \, P(y;\tau|y_1) \, dy \qquad (2.4.1)$$

is often used as the starting equation for the analysis of stationary stochastic processes and indeed for the theory of Brownian motion. Here $P(y_2; t \,|y_1)$ is the probability that the process undergoes a transition from the initial value y_1 to a final value y_2 at time τ through a sequence of intermediate values, i.e., the process goes from y_1 to y in the time τ, then from y to y_2 in the time interval $t - \tau$[7] and the intermediate values of the state variable are integrated over. Equation (2.4.1) was first introduced by Bachelier (1900) in his Ph.D. thesis (under the direction of Poincaré) on speculation in the stock market. As pointed out by Montroll and Shlesinger (1984), Bachelier earned the remarkable distinction of being the first to publish in the 20th century style of stochastic processes, and yet exerting absolutely no influence on the further development of the subject. That influence was exerted by the later (independent) investigations of Smoluchowski (1906), Chapman (1916), and Kolmogorov (1931). We therefore refer to (2.4.1) as the Bachelier-Smoluchowski-Chapman-Kolmogorov (BSCK) chain condition [see also Montroll and West (1979)].

When the process under consideration has translational invariance so that it is independent of the origin of the coordinate system

$$P(y_2, t \,|y_1) = P(y_2 - y_1; t) \qquad (2.4.2)$$

and when the range of the variable y_j is unbounded $(-\infty < y_j < \infty)$ then (2.4.1), takes the form

$$P(y_2 - y_1; t) = \int_{-\infty}^{\infty} P(y_2 - y; t - \tau) \, P(y - y_1; \tau) \, dy. \qquad (2.4.3)$$

Further, when moments of the probability density P exist and certain added properties are satisfied which will be described subsequently, the BSCK chain condition (2.4.3) will be shown to be equivalent to a partial differential equation, the Fokker-Planck equation for $P(y_2, t \,|y_1)$ (Fokker 1914, Planck 1917). We now proceed with the derivation of that equation. For those more comfortable with mathematics in a biological context we offer the derivation in Ricciardi [1977; cf. Sec. II] as an alternative.

We define the moments of the transition probability function $P(y_2; t \,|y_1)$ associated with the BSCK chain condition as follows

$$a_n(z,\Delta t) = \int (z-y)^n \, P(z;\Delta t \,|\, y) \, dy \; ; \quad n = 1, 2, \cdots \tag{2.4.4}$$

When the moments have the property that only a_1 and a_2, the first and second moments, are proportional to Δt, i.e., the time interval for the transition $y \rightarrow z$ to occur, as $\Delta t \rightarrow 0$ and that all other moments vanish more rapidly than linear in Δt in this limit, then it is possible to derive a differential equation, the Fokker-Planck equation, for $P(x;t \,|\, y)$ in terms of

$$A(z) = \lim_{\Delta t \to 0} \frac{a_1(z;\Delta t)}{\Delta t} \tag{2.4.5a}$$

and

$$B(z) = \lim_{\Delta t \to 0} \frac{a_2(z;\Delta t)}{\Delta t} \, . \tag{2.4.5b}$$

Following Wang and Uhlenbeck (1945) we consider the integral

$$I = \int R(y_2) \, dy_2 \, \frac{\partial}{\partial t} \, P(y_2;t \,|\, y_1) \tag{2.4.6}$$

where $R(y)$ is an arbitrary function which vanishes sufficiently rapidly as $|y| \rightarrow \infty$. Then from the discrete definition of a partial derivative (2.4.6) can be rewritten as

$$I = \int R(y_2) \, dy_2 \, \lim_{\Delta t \to 0} \frac{P(y_2; \, t+\Delta t \,|\, y_1) - P(y_2;t \,|\, y)}{\Delta t}$$

$$= \lim_{\Delta t \to 0} \frac{1}{\Delta t} \left\{ \iint R(y_2) P(y_2;t+\Delta t \,|\, y;t) \, P(y;t \,|\, y_1) dy dy_2 - \int R(y_2) P(y_2;t \,|\, y_1) dy_2 \right\} \tag{2.4.7}$$

Now we Taylor expand $R(y_2)$ in the neighborhood of $y_2 = y$ to obtain

$$R(y_2) = R(y) + (y_2-y) \, R'(y) + \frac{1}{2!} \, (y_2-y)^2 \, R''(y) + \cdots$$

where the primes denote derivations with respect to y. Hence, as $\Delta t \rightarrow 0$

$$\lim_{\Delta t \to 0} \frac{1}{\Delta t} \int R(y_2) \, P \, (y_2;t+\Delta t \,|\, y;t) \, dy_2$$

$$= R(y) + A(y)R'(y) + \frac{1}{2!} \, B(y) \, R''(y) \tag{2.4.8}$$

When (2.4.8) is substituted into (2.4.7) and an integration by parts is performed to reduce R' and R'', the resulting expression is

$$\int dy \, R(y) \left\{ \frac{\partial P}{\partial t} + \frac{\partial}{\partial y} \, (A(y)P) - \frac{1}{2} \, \frac{\partial^2}{\partial y^2} \, (B(y)P) \right\} = 0 \, . \tag{2.4.9}$$

Since the integral (2.4.9) vanishes for an arbitrary function $R(y)$, the quantity in the bracket must vanish. If we replace y_1 by y_0, the value of y when $t = 0$, we obtain the Fokker-Planck equation.

$$\frac{\partial P}{\partial t} = -\frac{\partial}{\partial y} \, [A(y)P] + \frac{1}{2} \, \frac{\partial^2}{\partial y^2} \, [B(y)P] \tag{2.4.10}$$

with the function $P(y;t\,|\,y_0)$ having the property

$$\lim_{t\to 0} P(y;t\,|\,y_0) \;=\; \delta(y-y_0) \quad . \tag{2.4.11}$$

The definition of the moments (2.4.4) along with (2.4.5b) implies that $B(y)$ is a non-negative function of y.

A number of examples for various choices of the average drift $A(y)$ and the state dependent diffusion coefficient $B(y)$ have been listed in the review article of Montroll and West (1979). An especially interesting choice is that corresponding to simple Brownian motion in which $A(y) = 0$ and $B(y) = 2D$, a constant, reducing (2.4.10) to the diffusion equation of Einstein

$$\frac{\partial P(y;t\,|\,y_0)}{\partial t} \;=\; D\;\frac{\partial^2}{\partial y^2}\;P(y;t\,|\,y_0) \quad . \tag{2.4.12}$$

The properly normalized solution to (2.4.12) with the initial delta function condition (2.4.11) is the Gauss distribution (2.1.22). This equation was the focal point of the Einstein (1905) theory of Brownian motion that was published five years after Bachelier's thesis had been successfully defended in Paris. Bachelier had also found the first passage time for the process and the probability distribution in the presence of an absorbing barrier.

In his discussion of the genetic advantage of a mutation (evolution of dominance) within a population R.A. Fisher (1922) constructed an equation of the form (2.4.12) to describe the evolution of the probability of a given gene frequency developing in successive generations of the population. Subsequently he introduced a genetic drift term giving rise to an equation of the general Fokker-Planck form (2.4.10). The interested student is referred to the excellent monograph by Fisher (1929) for a complete discussion of these concepts.

It is worth emphasizing that the Fokker-Planck equation (2.4.10) is linear in the probability density as of course it must be to yield the Gauss distribution as a solution. Here we see a way of interconnecting certain aspects of the preceding discussions. Consider the linear Langevin equation [cf. (2.3.10)]

$$\frac{dX}{dt} \;+\lambda X \;=\; f(t) \tag{2.4.13}$$

discussed in §2.3. In a short time interval Δt we can write (2.4.13) as the change in X between two nearby points in time

$$\Delta X \;=\; X(t+\Delta t)-X(t) \;=\; -\lambda X(t)\,\Delta t + \int_{t}^{t+\Delta t} f(t)\,dt \tag{2.4.14}$$

indicating in our recurrent example the change in profit over the time interval $(t,t+\Delta t))$. The deterministic change is directly proportional to Δt and the fluctuating change is integrated over this interval. The increment in X can be used to construct the moments of the process defined by (2.4.5). Denoting an average over an ensemble of realizations of the fluctuations $f(t)$ by a bracket $<\cdot\cdot\cdot>$, we define

$$A(x) = \lim_{\Delta t \to 0} \frac{<\Delta X>}{\Delta t} \tag{2.4.15a}$$

$$B(x) = \lim_{\Delta t \to 0} \frac{<(\Delta X)^2>}{\Delta t} . \tag{2.4.15b}$$

as the average drift and diffusion of our profit in the stock, resectively. To evaluate these averages we need to specify the statistical properties of $f(t)$. In §2.3 we argued that conditions existed under which $f(t)$ is a zero-centered Gauss process. Here we assume those conditions to be satisfied so that (2.3.12) specifies the statistical properties of $f(t)$. Thus, using (2.3.12a) in (2.4.14) we obtain

$$<\Delta X> = -\lambda X \Delta t \tag{2.4.16a}$$

and from (2.3.12b)

$$<(\Delta X)^2> = \lambda^2 X^2 (\Delta t)^2 + \int_t^{t+\Delta t} dt_1 \int_t^{t+\Delta t} dt_2 <f(t_1)f(t_2)>$$

$$= \lambda^2 X^2 (\Delta t)^2 + D \, \Delta t \tag{2.4.17}$$

so that (2.4.15a) and (2.4.15b) reduce to

$$A(x) = -\lambda x \, , \tag{2.4.18a}$$

$$B(x) = D \, . \tag{2.4.18b}$$

Substituting (2.4.18) into the Fokker-Planck equation (2.4.10) yields

$$\frac{\partial P}{\partial t} = \frac{\partial}{\partial x}(\lambda x P) + D \frac{\partial^2}{\partial x^2} P \tag{2.4.19}$$

which differs from the diffusion equation (2.4.12) by the presence of a linear "drift" term. Equation (2.4.19) describes an Ornstein-Uhlenbeck process (1930) and for the boundary condition that the probability density vanish as $|x| \to \infty$ and the initial condition $\lim_{t \to 0} P(x;t \,|\, x_0) = \delta(x - x_0)$; it has the solution

$$P(x;t \,|\, x_0) = \frac{1}{[2\pi\sigma^2(t)]^{1/2}} \exp\left\{-(x - x_0 \, e^{-\lambda t})^2/2\sigma^2(t)\right\} \tag{2.4.20}$$

where

$$\sigma^2(t) = \frac{D}{\lambda}(1 - e^{-2\lambda t}) \, . \tag{2.4.21}$$

Using (2.4.20) we can evaluate the average of x to obtain $x_o \equiv X(0)$

$$\overline{X}(t) = \int x P(x \mid x_0;t) \, dx = X(0) \, e^{-\lambda t} \tag{2.4.22}$$

as well as its variance

$$\overline{X^2} - \overline{X}^2 = \int x^2 \, P(x;t \mid x_0) dx - \overline{X}^2 = \sigma^2(t) \quad . \tag{2.4.23}$$

Thus (2.4.20) can be written

$$P(x;t \mid x_0) = \frac{1}{\sqrt{2\pi\sigma^2(t)}} \, \exp\left\{ -(x - \overline{X}(t))^2 / 2\sigma^2(t) \right\} \tag{2.4.24}$$

and we see that the linear regression term arising in the Gauss distribution in §2.2 is given here by the average value of the variable as it should. Also, denoting the equilibrium distribution as the infinite time value of (2.4.24):

$$P_{ss}(x) = \lim_{t \to \infty} P(x;t \mid x_0) \quad , \tag{2.4.25}$$

we see that (2.4.24) reduces to (2.3.18), as it should.

A more recent discussion of these topics in a biological context is given by Ricciardi (1977), who, among other things, is interested in diffusion models of neuronal firing, stochastic models of population growth and applications to genetics. Many of the mathematical details we have skimmed over here are contained in his monograph. The reader is cautioned that any similarity in notation between the present work and that of Ricciardi is purely coincidental.

We have now demonstrated two alternative ways of understanding the same process. In the previous section we developed a linear stochastic differential equation to describe the evolution of an observable in time. The solution to such an equation, given for example by (2.3.11), is not easily interpreted. In order to relate the solution to measurements one must calculate the appropriate moments. For the sake of discussion we made the simplifying assumption in our example that market conditions, i.e., the environment, gives rise to a Gauss distribution of fluctuations driving the profitability of our stock. In that way we were able to explicitlly calculate the first two moments of the solution. In the present section we have shown that under these conditions the Fokker-Planck equation provides a completely equivalent description of the evolution of the system. In fact if one specifies the trajectory of the solution in phase space by the delta function $\delta(x - X(t))$, where $X(t)$ is the dynamic solution to (2.4.13) say, then the probability density $P(x;t \mid x_0)$ is

$$P(x;t \mid x_0) = \, < \delta(x - X(t)) > \quad . \tag{2.4.26}$$

Here we see that the probability density is given by smoothing (averaging) over an ensemble of trajectories in phase space, each one of which is distinguished by a separate realization of $f(t)$, all starting from the same initial point $X(0) = x_0$.

The definition (2.4.26) is completely general and does not rely on any particular form of the dynamic equations, or any particular properties of the ensemble of fluctuations. Rather than the traditional derivation of the Fokker-Planck equation given above it is possible to start from (2.4.26) and derive a more general equation of evolution for the probability density which reduces to the Fokker-Planck equation under appropriate conditions; see e.g., West et al. (1979). Although such a derivation would lead us away from our main discussion it is important to realize that the evolution of the probability density is a conceptual alternative to the stochastic differential equation approach.

---

[1]If ξ_1 has a Gauss distribution and ξ_2 has a Gauss distribution then $\xi = \xi_1 + \xi_2$ has a Gauss distribution given by the convolution $P(\xi) = \int_{-\infty}^{\infty} P(\xi - \xi_1) P(\xi_1) d\xi_1$.

[2]We do not discuss the logic underlying the construction of the computer code since that would take us too far afield, but we view the program as an experiment from which we can gather valuable information in a controlled way.

[3]The properties of stable distributions are discussed in Section 3.2.

[4]It wasn't until the end of the sixteenth century that this technique was introduced into science by the astronomer Tycho Brahe (1582).

[5]We are well aware of the oversimpifications in these arguments, but since this is not intended as a tutorial in economics we hope that we have not violated anyone's sensitivities too severly.

[6]Note that the unobservables in this measurement are *not* operationally defined and by definition cannot be defined in the experiment. However there may be other measurements that select out one of the variables in $\mathbf{Y}(t)$ for study, in which case it would no longer be an unobservable.

[7]Note that the stationary property implies that the final state (y_2, t) is independent of the intermediate time τ.

3. THE IMPORTANCE OF BEING NONLINEAR

In the preceding section we have sketched some of the dominant styles of thought used in the understanding of complex events in natural philosophy. In our discussions we stressed the dependence of these schema on the notion of linearity. The motivation for these presentations was to uncover the often implicit and obscure dependence of these schema on linear ideas. We are now in a position to examine the limitations of some of these ideas in describing a number of familiar, but all too often ill-understood phenomena. We maintain that these processes can be understood by determining the proper nonlinear descriptions to replace the inadequate linear ones. As the title of this section indicates, we focus on the reasons why a nonlinear perspective is to be preferred over a linear one.

In this section we attempt to describe how the transition from a linear to a nonlinear world view has been forced on the physical sciences. This transition is still in process so what I will do in §3.1 is trace the historical (selected) development of the linear world view which presently dominates the physical sciences and show how it has proven to be inadequate for solving some of the fundamental problems in physics. One consequence of the linear world view is the *principle of superposition* which qualitatively states that a complex event can be segmented into a number of simple components in order to understand the components separately and then recombined back into an organized whole that can be understood in terms of the properties of the components. Nonlinear interactions preclude the recombining of the constitutive elements back into such an organized whole. In fact the existence of such entities as *solitons* and *dissipative structures* depend on the breaking of the symmetry implied by the principle of superposition. Generalizations of this concept to nonlinear systems using such techniques as the inverse scattering transform have been developed, but they are as yet only applicable under special conditions, e.g., for solitons. Thus the application of the superposition principle in its simplest form is almost never justified in natural philosophy, claims to the contrary notwithstanding.

In the first serious investigations of physics one is invariably struck by the existence of simple laws such as momentum conservation ($\sum_j m_j v_j = constant$; m_j is the mass and v_j the velocity of the jth particle) and energy conservation ($\frac{1}{2}\sum m_j v_j^2 = constant$ for *a free particle*). One also encounters many well defined linear concepts, such as the mobility $\mu, (v = \mu \cdot E)$, say for a conduction electron in a metal in the presence of an electric field E; the resistance R of a conductor through which a current I is passing resulting in a voltage drop V ($V = IR$), etc.

Simple relations are ubiquitous, e.g., the mean square displacement of a random walker after N steps is proportional to N; the vibrational density of states $\rho(\omega)$ of an E-dimensional crystal at low frequencies ω scales as ω^{E-1}, and so on. Everywhere exponents seem to be integers and quantities are well-defined. For many years scientists in the biological and behavioral sciences have viewed this situation as a curiosity since it was so unlike the situation in their own fields of study.

When one studies actual systems which are sufficiently complex, these simple relations may not be in evidence and it is often the case that the best description is in terms of probability distributions. The uncertainty in one's knowledge is reflected in the properties of the distribution function. For example, a Poisson distribution is characterized by its first moment, all higher moments being equal to it; whereas a Gauss distribution is characterized by its first two moments, all higher order moments being expressible in terms of them. However, when distributions have long enough tails, the first few moments will *not* characterize the distribution because they diverge. Distributions with *infinite* moments characterize processes described by non-integer exponents and surprises that run counter to our intuition. Integer exponents can usually be traced back to the analytic behavior of an appropriate function which can be expanded in a Taylor series. Non-integer exponents imply the presence of singularities and the breakdown of a Taylor series due to the divergence of a coefficient. *The main theme of §3.2 is that singularities and thus non-integer exponents arise in complex systems because they exhibit randomness on many scales.* In §3.2 we discuss the Lévy distribution which describes intermittent or clustered processes having mean first passage times and maxima moments which scale with a characteristic non-integer exponent μ. The Lévy parameter μ has been identified with the fractal (defined in §3.2) dimensionality for the process and can also be expressed in terms of random walk parameters.

There is of course a fairly long history of the appreciation of deterministic nonlinear effects in the biological and behavioral sciences. These nonlinearities have to do with the mechanisms of growth and saturation in many processes. In §3.3 we examine, briefly, the development of some of the nonlinear rate equations in biology, chemistry and genetics and examine the application of some of these concepts to the evolution of innovation in technology. It will become abundantly clear that the phenomenon of saturation, whatever the context, is nonlinear in character and thus is central to all growth processes having limited resources. Saturation is significant not just from the point of view of limiting the growth of a population in time through a nonlinear term in a rate equation, but also from the fundamental constraints imposed on a biological unit by physical consideration. For example the strength of a bone increases with cross sectional area, but its weight increases with volume, consequently at some stress a bone will break under its own weight. Therefore it is quite important to examine the scaling laws that arise in physiology because nature strives to be efficient in biological systems and these scaling relations manifest this efficiency.

The discussion of the more traditional ideas of nonlinear dynamics is initiated in §3.4 where I launch into a presentation of the *relaxation oscillator*. This particular example was chosen because it is familiar to scientists outside the physical sciences, although it was originally developed to describe the operation of a certain vacuum tube. Its inventor van der Pol, saw in the relaxation oscillator the hope of describing many of the oscillatory but not monochromatic processes in the biological, physical and behavioral sciences.

3.1 Physics; a Linear World View

Lord Rutherford, one of the experimentalists whose discoveries helped to form the quantum mechanical view of the microscopic world, once remarked to his students that, "All science is either physics or stamp collecting." This attitude was not limited to those working within physics and the other physical sciences, but has by and large come to be shared by many investigators in natural philosophy. Thus for better or worse, physics has become the paradigm against which most other scientific theories are gauged. These comparisons have very often been made explicit for a variety of reasons. However, it is not these explicit comparisons that are of concern here, since investigators usually go to great lengths to establish these connections. Of more immediate interest are the implicit assumptions that underlie the mathematical models in natural philosophy and which are due to the pervasive acceptance of the physics paradigm. We have mentioned a few of these concepts in our preceding discussions, e.g., the idea of isolating the system of interest from the environment. This can often be done in the physical sciences, but is almost completely unrealizable in the other realms of natural philosophy. The notion of isolating a process (system) is an intrinsically linear one, that is to say, the coupling of a process to the environment can only be made arbitrarily weak when the coupling is independent of the system. If the coupling is dependent on the process then for some values of the quantities characterizing the system the coupling to the background cannot be ignored. Such a system-dependent coupling is therefore nonlinear. For example in the derivation of the linear Langevin equation [cf. §2.3], the fluctuations $f(t)$ and dissipation $-\lambda X_0(t)$ arise from the linear coupling of the system to the environment. If this coupling between the system and environment were not linear then the fluctuations would depend on $X_0(t)$, i.e.,

$f(t) \rightarrow f(X_0(t), t)$, and the corresponding dissipation would be nonlinear. The particular form of the nonlinear dissipation would depend on the specific choice of the system-background coupling. [See e.g. Lindenberg and West (1984) and Cortes, West and Lindenberg (1985) where many of these ideas are discussed in detail.]

3.1.1 Linear Superposition

To understand the overwhelming importance of the linearity concept in the physical sciences we briefly review how the traditional style of thought used in physics is tied to linearity. We begin therefore with the first of the classical physicists, Sir Isaac Newton. We are *not* particularly interested in the form of his three laws of motion which are the foundation of mechanics, but rather we wish to gain some insight into his view of the physical world, i.e., how he applied these laws, and how that view influenced subsequent generations of scientists. Rather than presenting an extensive discourse on Newton's "Natural Philosophy" we examine how he answered one of the outstanding physics questions of his day. The problem was to determine the speed of sound in air. Our focus will be on how Newton the physicist thought about difficult problems for which he had no exact solution, or indeed no equation of motion, and how the success that emerged from his style of thought or strategy of model construction motivated future generations of scientists to think in the same way.

Recall that sound is a wave phenomenon and its description is given in elementary physics texts by a wave equation. However the wave equation is a partial differential equation and such mathematical expressions had not been invented at the time Newton considered this problem (1686). He did know however that sound was a wave phenomenon because of the observations of such effects as reflection, refraction, and diffraction. Although the proper mathematics did not exist, Newton argued that a standing column of air could be modeled as a linear harmonic chain of equal mass oscillators. The picture consisted of air molecules of equal masses arranged along a line and locally interacting with other air molecules by means of a linear elastic force to keep the molecules localized. The system of coupled oscillators was a natural model for the inventor of the calculus (fluxions) to develop because it required individual air molecules (oscillators) to undergo periodic motion during the passage of a disturbance. A sound wave, after all, consists of periodic rarefactions and compressions in the air and is an ideal physical system for such a model.

The linear character of the model enabled Newton to reason to the solution of the equation of motion without explicitly writing it out. He was able to deduce that the speed of sound in air is $v = (p/\rho)^{1/2}$ where p is the pressure of the wave and ρ is the mass density of the column of air. Using the isothermal volume elasticity of air, i.e., the pressure itself, in his equation he obtained a speed of sound in air of 945 feet/second, a value 17 percent smaller than the observed value of 1142 feet/second.

Newton's argument was severely criticized by Lagrange (1759) who presented a much more rigorous statement of the problem. Lagrange developed a simple mechanical model of an equivalent system, that of a string, which he represented by a set of ordinary differential equations, one for each spring. If ξ_n represents the displacement of the $n\,th$ oscillator from its equilibrium position, the equations he constructed are

$$\frac{d^2\xi_n}{dt^2} = \alpha(\xi_{n+1} - \xi_n) - \alpha(\xi_n - \xi_{n-1}) \; ; \; n = 1 \; .... \; N \; . \tag{3.1.1}$$

The n-oscillator is then linearly coupled to the (n $\pm$ 1) - oscillators, i.e., its nearest neighbors, and the force causing this oscillator to move is proportional to the net displacement of these three oscillators from their rest positions. A crucial assumption in (3.1.1) is that the force only acts locally along the chain, i.e., the $n\,th$ molecule moves in response to the activity of its nearest neighbors. This happens all down the chain described by the differential-difference equation (3.1.1) , i.e., an equation that is continuous in time t but discretely indexed by its position along the chain n.

As Lindsay (1945) points out in the Historical Introduction to Rayleigh's *Theory of Sound*, Lagrange must have been quite surprised when his rigorous derivation of the speed of sound gave Newton's result. That is how the matter stood until 1816 when Laplace (1825) replaced the isothermal with the adiabatic elastic constant, arguing that the compressions and rarefactions of a longitudinal sound wave take place adiabatically. He obtained essentially exact agreement (within certain small error bounds) with experiment. Thus, the modeling of the complicated *continuum* process of sound propagation in air by a discrete linear harmonic chain was successful! *The success of the application of a harmonic chain to the description of a complicated physical process established a precedent that has evolved into the backbone of modeling in physics.*

The next steps in the formation of a linear world view were taken largely by a father and son team through the letter correspondence of John (father) and Daniel (son) Bernoulli beginning in 1727. In their correspondence they established that a system of N point masses has N independent modes of vibration, i.e., eigenfunctions and eigenfrequencies. The number of component independent degrees of freedom of the system motion is equal to the number of entities being coupled together. The Bernoulli's reasoned that if you have N point masses then you can have N modes of vibration without making particular assumptions about how to construct the equations of motion as long as the coupling is weak. This was the first statement of the use of eigenvalues and eigenfunctions in physics. Simply stated, their conclusion means that one can take a complicated system consisting of N particles or N degrees of freedom and completely describe it by specifying one function per degree of freedom. If one knows what the eigenfunctions are (for waves these are harmonic functions) and one knows what the eigenvalues are (again for waves these are the frequencies) then one knows what the motion is. Thus, they had taken a complicated system and broken it up into readily identifiable and tractable pieces, each one being specified by an eigenfunction. If one knows the eigenfunctions and eigenvalues then the contention was that one knows nearly everything about the system.

Later, Daniel (1755) used this result to formulate the *principle of superposition*: the most general motion of a vibrating system is given by a linear superposition of its characteristic (eigen, proper) modes. The importance of this principle cannot be over emphasized. As pointed out by Brillouin (1946), until this time all general statements in physics were concerned with mechanics, which is applicable to the dynamics of an individual particle. The principle of superposition was the first formulation of a *general law* pertaining to a system of particles. One might therefore date theoretical physics from the formulation of this principle. Note that the concepts of energy and momentum were still rather fragile things and their conservation laws had not yet been enunciated in 1755. Yet at that time Daniel Bernoulli was able to formulate this notion that the most general motion of a complicated system of particles is nothing more than a linear superposition of the motions of the constituent elements. That is a very powerful and pervasive point of view about how one can understand the evolution of a complex physical system.

This procedure was in fact used by Lagrange (1759) in his study of the vibration of a string. Using the model (3.1.1) in which there are a finite number of equally spaced identical mass particles, he established the existence of a number of independent frequencies equal to the number of particles. This discrete system was then used to obtain the continuous wave equation by letting the number of particles become infinitely great while letting the masses of the individual particles vanish in such a way that the mass density of the string remained constant. The resulting frequencies of the continuum were found to be the harmonic frequencies of the stretched spring.

During the same period the most prolific theoretical physicist of that age, or any age, Euler, was investigating the vibrations of a continuous string (1748). In his analysis he used a partial differential equation, as had Taylor some 30 years earlier, and was able to demonstrate for the first time that the transverse displacement of the string during a vibration can be expressed as an arbitrary function of $x \pm vt$, where x is the distance along the string, t is the time and v is the velocity of the vibration. Euler's solution had the form

$$u(x,t) = \sum_{n=1}^{\infty} \hat{f}(n)\cos n\pi t \sin n\pi x \qquad (3.1.2)$$

where the initial disturbance $f(x)$ has the form

$$f(x) = \sum_{n=1}^{\infty} \hat{f}(n) \sin n\pi x \quad . \qquad (3.1.3)$$

A similar result had been obtained by d'Alembert the preceding year (1747) in his discussion of the oscillation of a violin string. His (d'Alembert) proposed solution had the form

$$u(x,t) = \frac{1}{2} f(x+vt) + \frac{1}{2} f(x-vt) \quad . \qquad (3.1.4)$$

which can also be obtained from (3.1.2) by means of trigonometric identities for sines and cosines. Euler and d'Alembert introduced partial differential equations into the discussion of physical processes, in particular they studied solutions of the wave equation

$$\frac{\partial^2}{\partial t^2} \xi(x,t) = v^2 \frac{\partial^2}{\partial x^2} \xi(x,t) \quad . \qquad (3.1.5)$$

This style of thought provided the theoretical underpinning for D. Bernoulli's (1755) insight into the principle of superposition.

Oddly enough Euler did not accept the principle of superposition for the following reason. Since it was known that the characteristic modes of vibration were sine and cosine functions and since the general motion of the string was a function only of $x \pm vt$, it followed that the most general vibration of the string could be expressed as a series (superposition) of sines and cosines with arguments $x \pm vt$. Euler thought that superposition was fine as long as one were talking about individual waves that were real physical objects -- say the waves on a transverse string. In other words, if one has a string and it is juggled at one end, waves begin to travel up and down the string and these can be denoted by sines and cosines. Thus one can take a superposition of them to give the overall displacement of the string. However, he thought, the notion of writing the *general motion* of a physical system as a superposition, the way D. Bernoulli contended, was unphysical. He did not think you could write down the general motion of a system in that way, even though he was the first to show that you could write down the solutions to the wave equation that way. As Brillouin (1946) points out, this was almost an absurdity to Euler's mind. Since he did not doubt his own results (3.1.3), he rejected the principle of superposition. It turns out 230 years later that he was not entirely wrong.

You will no doubt recognize the above statement of the general motion (3.1.2) as a special case of Fouriers' theorem: *any periodic function can be expressed as a series of sines and cosines of some fundamental period T*;

$$f(t) = \sum_{n=0}^{\infty} \left\{ a_n \cos\left(\frac{2\pi nt}{T}\right) + b_n \sin\left(\frac{2\pi nt}{T}\right) \right\} \tag{3.1.6}$$

with the coefficients of the individual harmonic terms determined by

$$a_n = \frac{1}{T} \int_{-T}^{T} f(t) \cos\left(\frac{2\pi nt}{T}\right) dt \tag{3.1.7a}$$

$$b_n = \frac{1}{T} \int_{-T}^{T} f(t) \sin\left(\frac{2\pi nt}{T}\right) dt . \tag{3.1.7b}$$

This idea arises naturally in astronomy; in fact Neugebaurer (1952) discovered that the Babylonians used a primitive kind of Fourier series for the prediction of celestial events.

Fourier's own contribution to the linear world view began in 1807 in his study of the flow of heat in a solid body. He established the principle of superposition for the heat equation. Note that the heat equation (to be discussed in some detail shortly) describes irreversible, dissipative processes unlike the wave equation which describes reversible periodic processes. It is significant that the principle of superposition could be applied to these two fundamentally different physical situations. Fourier (1822) argued, and it was eventually proved rigorously by Dirichlet (1829), that any piecewise continuous (smooth) function could be expanded as a trigonometric (Fourier) series, i.e., as sums of sines and cosines with proper weightings. Thus, the emerging disciplines of acoustics, heat transport, electromagnetic theory (Maxwell's equations) in the eighteenth and nineteenth centuries could all be treated in a unified way. The linear nature of all these processes insured that they satisfy the principle of superposition.

The penultimate form of the principle of superposition was expressed in the mathematical formulation of the Sturm-Liouville theory (1836-1838). The authors of this theory demonstrated that a large class of differential equations (and thereby the associated physical problems intended to be modeled by these equations) could be cast in the form of an eigenvalue problem. The solution to the original equation could then be represented as a linear superposition of the characteristic or eigen-motions of the physical problem. This was immediately perceived as the systematic way of unraveling complex events in the physical world and giving them relatively simple mathematical descriptions. In the following century this would allow the problems of the microscopic world to be treated in a mathematically rigorous way. All of quantum mechanics would become accessible due to the existence of this mathematical apparatus.

Thus at the turn of the century it was the commonly held belief in the physical sciences community that: *a complex process can be decomposed into constituent elements, each element can be studied individually and the interactive part deduced from reassembling the linear pieces.* This view formed the basis of macroscopic field theories and rested on the twin pillars of the Fourier decomposition of analytic functions and Sturm-Liouville theory. Physical reality could therefore be segmented; understood piece-wise and superposed back again to form a new representation of the original system. This approach was to provide the context in which other physical problems were addressed. Moreover, all deviations from linearity in the interacting motions were assumed to be treatable by perturbation theory. Linear theory describes the dominant interaction, and if there are any remaining pieces, e.g., the interactions are not just nearest neighbors in a linear chain, but involve next nearest neighbors or perhaps a nonlinear interaction, it is assumed that the strengths of these remaindering interactions are small. Thus the philosophy was to solve the linear problem first, then treat the remaining interaction (that was not treated quite properly) as a perturbation on the linear solution, assuming throughout that the perturbation is not going to modify things significantly. This is the whole notion of interpolating from a linear world view to the effects of nonlinearity. In the linear world view, all the effects of nonlinearity are perturbative (weak) effects.

3.1.2 Irreversibility

Physical phenomena such as turbulence (chaotic fluid flow) and concepts such as irreversibility were not in the mainstream of physics after the turn of the century, but almost everyone was confident that some clever use of perturbation theory would properly account for the deviations from linearity manifest in such contrary processes. *So what is wrong with the linear world view?* After all, even the infinities of quantum field theory had eventually yielded to a properly formulated renormalized perturbation representation. This tightly woven point of view began to unravel when it was carefully applied to the understanding of macroscopic irreversible phenomena. The scientists E.Fermi, S. Ulam and J. Pasta (1955) began an investigation into the irreversible process of heat transport in solids in a way analogous to Newton's calculation of the velocity of sound in air. Before presenting their argument let us review the process of simple heat transport in a solid to understand the exact nature of the problem.

It has been known since the work of Fourier mentioned earlier that the conduction of heat is governed by the phenomenological equation

$$\partial_t \, T(x,t) \; = \; \kappa \, \partial_{xx} \, T(x,t) \tag{3.1.8}$$

(in one dimension, e.g., along a wire) where $\partial_t \equiv \dfrac{\partial}{\partial t}$, $\partial_{xx} \equiv \dfrac{\partial^2}{\partial x^2}$, $T(x,t)$ is the macroscopic temperature of the material at time t and location x, and κ in the heat diffusivity. The heat equation is solved as a boundary value problem; we specify the value of $T(x,t)$ at the ends of the wire as $T=0$ at $x=0$ and $T=T_0$ at $x=L$ for all times, say. If the temperature of the wire is zero initially then the solution to Fourier's heat equation is:

$$\frac{T(x,t)}{T_0} \; = \; \frac{x}{L} + \sum_{l=0}^{\infty} \frac{(-1)^l}{l\pi} \, \sin\left[\frac{l\pi x}{L}\right] \exp\left\{ -\kappa \, t \, \left[\frac{l\pi}{L}\right]^2 \right\} \tag{3.1.9}$$

from which we observe the superposition nature of the solution. As $t \to \infty$ the coefficients in the series go to zero since $\kappa(l\pi/2)^2 > 0$, leaving a linear change of temperature from 0 to T_0 as the steady state temperature profile of the wire. This latter fact is the point I wish to emphasize by writing down this solution, that the process is irreversible, i.e., one cannot reverse the direction of time in the solution as one can in a simple mechanics problem and regain the initial state of the system.

Let us now return our attention to the question of irreversibility as it arises in the heat equation. Firstly, we note that the dynamic microscopic equations that underlie the macroscopic heat transport equation are reversible in time. A physicist would postulate that there exists a Hamiltonian H to describe the degrees of freedom in the metal. The simplest physical picture is one in which the constituent parts of the metal are described by a vector of positions $\mathbf{q}(t)$ and a vector of canonically conjugate momenta $\mathbf{p}(t)$ and whose equations of motion are:

$$\dot{\mathbf{q}}(t) \; = \; \frac{\partial H}{\partial \mathbf{p}} \quad ; \quad \dot{\mathbf{p}}(t) \; = \; -\frac{\partial H}{\partial \mathbf{q}} \tag{3.1.10}$$

where we need not as yet specify the form of the Hamiltonian, or in particular the interaction among the atoms in the metal. In the absence of a magnetic field if we reverse the direction of time in (3.1.10), i.e., $t \to -t$, then $\mathbf{q}(t) \to \mathbf{q}(-t)$ and $\mathbf{p}(t) \to -\mathbf{p}(-t)$ and the equations of motion remain the same. Thus the fundamental question arises: *How can one construct a microscopic model of an irreversible process, such as described macroscopically by (3.1.8) when the microscopic equations (3.1.10) are reversible?*

There have been a number of elaborate attempts to answer this reversible-irreversible question but none has been completely successful. It is often the case that it is easier to address a general question than to answer any specific one. Thus, despite all these various attempts at describing irreversibility no one has succeeded in providing a microscopic description of the Fourier law of heat transport. This state of affairs motivated the study that has in part stimulated the present resurgence of interest in nonlinear dynamics that has exploded into the physics and mathematics communities.

The investigation initiated by Ulam, Fermi and Pasta (1955) at the Los Alamos National Laboratory was *intended* to show how macroscopic irreversibility resulted from the deviation of the microscopic interactions from linearity. Although they did not succeed in this attempt they made a discovery of perhaps equal importance. They considered a mathematical model of a solid based on the harmonic approximation in which the solid is considered to consist of a network of coupled harmonic oscillators in the spirit of Newton's calculation of the speed of sound in air. This three dimensional harmonic lattice is adequate for the description of many physical properties including the spectrum of vibrations in a metal. A localized excitation of the metal will remain coherent since the harmonic lattice consists of a discrete spectrum with a *finite* number of degrees of freedom. If a normal mode of the material is excited, the energy will remain in that mode forever in the harmonic approximation, see e.g., Maradudin et al. (1971). This description is at variance with one of the fundamental precepts of classical statistical mechanics: *for a given temperature of the environment the system will equilibrate in such a way that each degree of freedom (normal mode) has the same amount of energy, i.e.,* $\frac{1}{2}k_B T$. This is the equi-partition theorem and it is violated by a strictly harmonic lattice. It was shown by Ford, Kac and Mazur (1965) that the equi-partition theorem does apply if one investigates the excitation of a simple oscillator in the limit of an *infinite* number of degrees of freedom in the solid.

Fermi, Pasta and Ulam reasoned that for equi-partitioning to occur, that is, for a heat pulse, say, to equilibrate in the metal by raising the overall temperature slightly, a nonlinear interaction mechanism must be present. They therefore included a weak anharmonicity in each of the oscillators and anticipated that a calculation of the dynamics would support the principle of energy equi-partition. The average strength of the nonlinearity is given by the parameter λ which is assumed to be an order of magnitude smaller than that of the linear parameter α. The equations of motion are

$$\dot{q}_n = p_n$$

$$\dot{p}_n = \alpha(q_{n+1} - 2q_n + q_{n-1}) + \lambda[(q_{n+1} - q_n)^2 - (q_n - q_{n-1})^2] \qquad (3.1.11)$$

The equi-partition theorem, however, is deduced from equilibrium statistical mechanics and not from a determination of the lattice dynamics although it is always argued that the two descriptions are equivalent. Therefore, they were quite surprised when equilibrium was not achieved by the anharmonic lattice, that is to say there was very little energy sharing among the modes of vibration on the average.

The displacement of the n th oscillator in the normal mode representation is given by the discrete *Fourier series* of N modes,

$$q_n(t) = \frac{2}{N} \sum_{k=1}^{N} a_k(t) \sin(nk\pi/N) \qquad (3.1.12)$$

where $a_k(t)$ is the k th mode amplitude. For $\lambda = 0$ in (3.1.11) the normal modes satisfy the N equations for independent linear harmonic oscillators

$$\ddot{a}_k(t) + \omega_k^2 a_k(t) = 0 \qquad (3.1.13)$$

where $\omega_k = 2\sin(\pi k/2N)$ and the total mechanical energy of the solid is

$$E = \sum_k \left\{ \frac{1}{2} \dot{a}_k^2 + 2a_k^2 \sin^2(\pi k/2N) \right\} . \qquad (3.1.14)$$

Since the system is conservative, i.e., there is no dissipation, the total energy E is constant. In the linear system if all the energy were initially put into a single mode, mode 1 say, then only $a_1(t)$ would oscillate since it is uncoupled from the rest of the system, i.e., there is no coupling among the modes in (3.1.13). The nonlinear terms however couple the linear eigenmodes together, that is to say, energy is transferred from one eigenmode to another by the nonlinear interactions. The chastity of linear superposition is thus violated by the presence of nonlinear interactions.

Thus the energy is transferred out of mode 1 into the neighboring modes by means of mode-mode coupling in the nonlinear interaction. For those interested the nonlinear equations have the normal mode form

$$\ddot{a}_k + \omega_k^2 a_k = \frac{8i\lambda}{N} \sum_k{}' a_k' a_{k-k}' \sin(\pi k/N) \sin(\pi k'/N) \sin(k'-k)\pi/N$$

$$= \sum_k{}' \Gamma_{kk}' a_k' a_{k-k}' = g_k(t) \qquad (3.1.15)$$

with the formal solution

$$a_k(t) = a_k(0) \cos \omega_k t + \frac{\dot{a}_k(0)}{\omega_k} \sin \omega_k t$$

$$+ \int_0^t g_k(\tau) \sin \omega_k (t - \tau) d\tau \tag{3.1.16}$$

which is a nonlinear integral equation in the a's and $\{a_k(0), \dot{a}_k(0)\}$ are the set of initial conditions for the mode amplitudes and their time derivatives. Now even though energy has only been put into a single mode initially, in the equation driving the other modes, this energetic degree of freedom provides a source of energy for the other modes. This energy is pumped out of mode 1 into mode 2 and mode 2 begins to grow. Then while mode 2 is growing, energy is also supplied to mode 3 and it begins to grow at a somewhat slower rate than mode 2 [cf. (Figure 3.1.1)]. Fermi, Pasta and Ulam used 2^4, 2^5 and 2^6 modes in the computer simulation of the set of equations (3.1.15). In their calculations a significant amount of energy never got beyond the fifth mode. As a matter of fact, when they waited long state reconstituted itself, i.e., all the energy was recollected into mode 1. Rather than equi-partitioning with all the energy uniformly distributed among the available N degrees of freedom, what occurs is that the initial state of the system periodically re-emerges from the dynamics. The time interval between such visits is now called the FPU recurrence time. Thus there is no energy equi-partitioning, no thermodynamic equilibrium and there is certainly no explanation of the irreversibility that is manifest by the heat equation.

We now come upon a case of history repreating itself. Newton in 1686 constructed a discrete mathematical representation of linear waves and Euler in 1748 did it for a continuous string; the two were essentially physically equivalent and the properties of the solution were the same (ignoring the difference between longitudinal and transverse waves for now). Fermi, Pasta and Ulam in 1955 constructed a discrete nonlinear wave system and in 1965 Zabusky and Kruskal, essentially constructed the continuous form of the discrete set of equations used by FPU. Zabusky and Kruskal (1965) considered the case when the lattice spacing in the oscillator system vanishes $\Delta \to 0$. This converts the FPU difference-differential equations into a nonlinear partial differential equation:

$$\frac{1}{v^2} \frac{\partial^2}{\partial t^2} \xi(x,t) = \left(1 + \lambda \frac{\partial \xi}{\partial x}\right) \frac{\partial^2}{\partial x^2} \xi + \frac{\Delta^2}{12} \frac{\partial^4}{\partial x^4} \xi + O(\Delta^3) \quad . \tag{3.1.17}$$

This equation has solutions $\xi(x,t) = f(x \pm vt)$ just as does the wave equation, except that the present solutions are called solitons. They are localized propagating waves and are the "normal modes" of the nonlinear equation of motion. Soliton solutions have the properties:

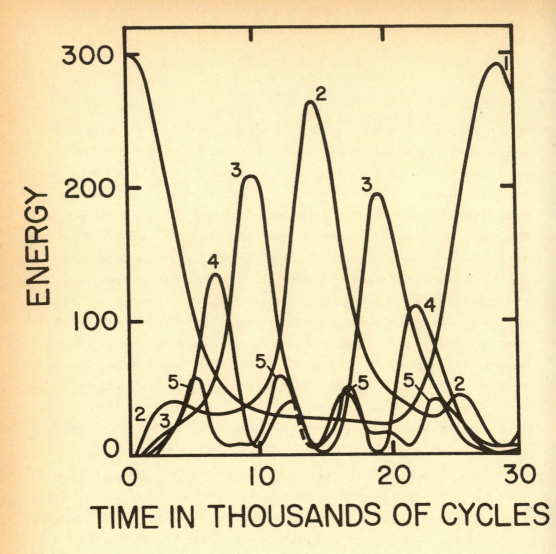

Figure 3.1.1.   The variation in energy of various modes as a function of time. The units of energy are arbitrary, the coupling coefficient is 0.25 and the total number of modes is 32. Modes higher than the 5th never exceeded 20 units of energy on the scale given here [Fermi, Pasta and Ulam (1955)].

1.　　solitons do not become diffusive as they propagate even though the medium supporting the motion must be dispersive;

2.　　solitons are stable against interactions with other solitons;

3.　　an arbitrary finite amplitude waveform will breakup into one or more stable solitons as well as into non-soliton wave forms.

See e.g., Ablowitz and Segur (1981) for a complete discussion of solitons using the inverse scattering transform technique.

Let us now reconsider the FPU calculation. If the initial state of the anharmonic chain is a soliton then this corresponds to a certain set of values of the linear mode amplitudes. When the soliton is reflected from the chain ends, it returns to its initial configuration. The energy flowing from one normal mode into another reverses direction in this case so that the initial distribution of energy among the normal mode amplitudes repeats itself. Thus the picture that unfolds is one in which the initial state of the chain is close to a soliton state which preserves its character for a long time (a true soliton would last forever). This makes it seem that the anharmonic chain is not ergodic and that energy gets transferred in and out of the lower modes periodically. Nonlinearities can therefore provide *coherent structures* as well as the incoherent state of thermal equilibrium.

So what is the general world view that emerges from this discussion? In simplest form we consider the principle of superposition and write the solution to a linear dispersive field equation as

$$u(x,t) = \frac{1}{2\pi} \int_{-\infty}^{\infty} e^{i[kx - \omega(k)t]} \; \tilde{u}\left(k,\omega(k)\right) dk \quad . \tag{3.1.18}$$

A dispersion relation gives the frequency as a function of wavenumber thereby relating the spatial with the temporal behavior of the field. For a narrow band process centered on $k = k_0$ one can show that

$$u(x,t) \sim \frac{\exp\{i[k_0 x - \omega(k_0)t]\}}{\sqrt{t}} \tag{3.1.19}$$

using the method of stationary phase so that the field amplitude decays as $t^{-1/2}$. The point of this presentation is that if one has a linear dispersive wave field and energy is injected as a pulse with central wavenumber k_0, then asymptotically, no matter how it moves around, because the field is linear the wave amplitude will decay as $t^{-1/2}$. If one starts with a bump, it will become flat no matter what one does.

If the wave field has the proper nonlinearity -- and here comes the change in viewpoint -- one can start out with a bump at $t=0$, and under certain conditions one can have the same bump at $t = \infty$. The strength of the nonlinear terms in the field equations is such as to exactly balance the effect of linear dispersion. This delicate balance of linear dispersion driving the component waves apart and nonlinear interactions pulling them together provides for coherent structures such as solitons and generally inhibits the tendency of systems to separate into their incoherent linear components. Thus the physical effects of linear dispersion and nonlinear interactions can reach a dynamic balance to form a soliton. The impact of this phenomenon is not limited to physics alone, e.g., there is strong evidence for the existence of quantum mechanica solitons in macromolecules [see e.g., Scott et al. (1983)] that may in part provide the mechanism for energy transport in biological systems.

So what about irreversibility in the heat equation? We started from a mathematical model of an anharmonic solid and found that instead of an irreversible approach to a state of energy equi-partition as predicted by Fourier's heat law, we obtained coherent solitons. This unsettling result has stimulated a flood of interest in the nonlinear alternatives to the linear world view outlined above. The microscopic derivation of the heat law has not as yet been obtained, but a much greater understanding of the properties of nonlinear systems such as (3.1.15) has been developed since the seminal FPU study. A presentation of the relevant conceptual developments in nonlinear dynamics and non-Gauss statistics as they may be used in natural philosophy forms the remaining sections of this essay.

3.2 Some Non-Gauss Statistics, Clustering and Fractals

We have now seen how the concept of linearity is inconsistent with the existence of a soliton. It had been hoped that the breaking of the linearity assumption would lead to a "more fundamental" result, that being the equi-partitioning of the energy in a system having a significant number of degrees of freedom, but FPU instead observed the formation of a dynamic coherent state. We mentioned that the soliton persists because of a balancing between the linear dispersion of the system attempting to degrade the wave pulse and the nonlinear interaction attracting the pulse towards its central frequency. This dynamic balance is one kind of mechanism that is intrinsically nonlinear and could not have been predicted from any extrapolation of the linear world view. Without a nonlinear interaction there would be no soliton. Thus the thermodynamic property that FPU were attempting to explain remained unexplained; the deterministic interaction mechanism was not sufficient to decipher the statistical result of energy equi-partition. We will return to this topic subsequently, but first we examine some stochastic processes that although quite common violate the short term linear assumption made for Gauss statistics. In fact we will find that one of the most frequently occurring distributions in economics and elsewhere, the Pareto distribution, is a consequence of a nonlinear amplification process. But this is getting too far ahead of our story.

Recently, considerable attention has been devoted to diverse phenomena exhibiting a clustered behavior in the space and/or time domains as well as in the more general clustering in the phase space of the system. Examples of such clustered behavior appear in data from the physical sciences in such varied processes as the intermittency in turbulent fluid flow [see e.g., Frisch, Sulem and Nelkin (1978)]; the distribution of errors in data transmission system [Mandelbrot (1977), Chap. IV], the clustering of galaxies [Peebles (1980)] to name a few; also in the medical literature there are a number of examples, for instance the distribution of the bronchial sizes in the lungs [Weibel (1963) also West, Goldberger and Bhargarva (1985)]. The spatial distribution of the conduction filaments in the His-Purkinje region of the heart [Bellet (1971)] and membrane noise [see e.g., Verveen and De Felice (1974)]. There is apparently a close connection between such clustered behavior in space or time and the Hausdorff-Besicovitch dimensionality of these processes. In his two extended essays Mandelbrot (1977, 1982) has introduced the term fractal into the scientist's lexicon along with the geometric notion of a fractal dimension. For most purposes the fractal and Hausdorff-Besicovitch dimension are interchangeable, so that keeping with modern fashion we will use the shorter term, fractal dimension.

The fractal dimensionality D of a set may be defined as follows. If a finite set Ω is divided into N identical parts, each of which is geometrically similar to Ω with the similarity ratio r, then $D = \ln N/\ln(1/r)$, where $r < 1$ is a real positive number. Analytic estimates of the fractal dimensionality of dynamical processes have so far been carried out only for Brownian motion or the more general Lévy motion, although numerical estimates for a number of systems have been made. A simple measure of the fractal dimensionality of a model dynamical process that exhibits clustering has been developed by Hughes, Shlesinger and Montroll (1981). Their model is a discrete random walk on a lattice with transition probabilities drawn from a distribution having long range interactions on the lattice. They were able to associate the short-scale behavior of the lattice structure function to the fractal dimensionality of the walk. The continuum limit of their random walk model lead to the equation of evolution for a Lévy distribution. In this section we discuss the properties of this probability distribution and find that it is the natural generalization of the distribution of Gauss and includes the latter as a special case. Among its varied and interesting properties is the fact that it obeys a scaling law, indicating that the statistical fluctuations in the underlying process maintain a self-similarity property. This is not completely unexpected since the random walk model on which this equation is based has this scaling feature built in, and in fact is referred to as a fractal random walk. Let us now examine the properties of a Lévy process.

3.2.1 Lévy Processes

The derivation of the Fokker-Planck equation (2.4.10) requires that the first two moments (3.1.5) of the state variable exist, as well as certain other properties of the higher moments of the dynamic process $X(t)$. Certain distributions such as that of Cauchy:

$$P_c(x_2 - x_1; t) = \frac{at}{\pi[(x_2 - x_1)^2 + a^2 t^2]} \tag{3.2.1}$$

satisfy the BSCK chain condition (3.1.1) but *do not have the bounded moments required for the derivation of* (2.4.10). These cases have been investigated by Lévy (1925) and are discussed in this subsection. We limit our discussion to the translationally invariant distributions characterized by $P(x_2; t \mid x_1) = P(x_2 - x_1; t)$ which since they satisfy the BSCK chain condition are also stationary Markov processes. For the purposes here we interpret the Markov property to mean that an n-point probability density $P(x_1, t_1; x_2, t_2; \cdots; x_n, t_n)$ can be expressed as a product of $(n-1)$-two-point probability densities $P(x_1; t_1 \mid x_2; t_2) P(x_2; t_2 \mid x_3; t_3) \ldots P(x_{n-1}; t_{n-1} \mid x_n; t_n)$. We stress that the class of distributions we discuss here, of which the Cauchy P_c is but one example, has the distinguishing feature of having a *power-law form* for large values of the variable. For example in (3.2.1) when $|x_2 - x_1| \gg at$ the second term in the denominator can be neglected with respect to the first and the distribution falls off as the square of the variable. It is this quadratic dependence of the probability density that gives rise to the result

$$\int_{-\infty}^{\infty} P_c(x; t) \, x^2 dx = \infty \quad,$$

i.e., the variance of the Cauchy distribution is infinite. This divergence of the central moments is representative of distributions having power-law tails. If the power-law has an exponent $(\alpha + 1)$ then all central moments $< |x|^p >$ where $p < \alpha$ will be finite and those with $p \geq \alpha$ will diverge.

We find it convenient to introduce the characteristic function $\phi(k, t)$ which is the Fourier transform of the probability density:

$$\phi(k, t) = \int_{-\infty}^{\infty} dx \, e^{ikx} P(x, t) \quad. \tag{3.2.2}$$

For processes satisfying the BSCK chain condition the characteristic functions satisfy the product rule

$$\phi(k,t) = \phi(k;t-\tau)\,\phi(k,\tau) \quad . \tag{3.2.3}$$

The Fourier transform of the Gauss distribution (2.4.24) for $x_0 = 0$ and $t \geq 0$

$$\phi(k,t) = e^{-Dtk^2} \tag{3.2.4}$$

satisfies (3.2.3). Hence the Gauss distribution obeys the BSCK chain condition. Another function which satisfies (3.2.3) is

$$\phi(k,t) = e^{-at|k|} \tag{3.2.5}$$

whose inverse Fourier transform is the Cauchy distribution (3.2.1). Indeed there are many other examples of functions which: (1.) satisfy the product rule (3.2.3); (2.) have Fourier transforms which are non-negative; (3.) are normalized so that the probability function $P(x,t)$ conserves probability at all $t \geq 0$.

The most general form of $\phi(k,t)$ satisfying the above three conditions was obtained by Lévy (1937) and by Khinchine and Lévy (1936). For symmetric processes, the most general form is given by

$$\phi(k,t) \equiv e^{-bt|k|^\alpha} \tag{3.2.6}$$

where b is a positive real[1] constant and $0 < \alpha \leq 2$. Clearly the corresponding probability distribution

$$P_\alpha(x,t) = \frac{1}{2\pi} \int_{-\infty}^{\infty} e^{-bt|k|^\alpha} e^{-ikx}\, dk \tag{3.2.7}$$

is normalized since

$$\int_{-\infty}^{\infty} dx\, P_\alpha(x,t) = \int_{-\infty}^{\infty} \delta(k) e^{-bt|k|^\alpha}\, dk = 1 \quad . \tag{3.2.8}$$

The distributions characterized by (3.2.7) were first proposed by Cauchy (1853) but he did not investigate the non-negativity in detail nor did he note the restriction $0 < \alpha \leq 2$.

The class of processes whose characteristic functions satisfy (3.2.7) are called Lévy processes. Many well known stochastic processes of physical interest can be described by a Lévy distribution as we discuss shortly. We mentioned that the choice of parameter $\alpha = 2$ correspond to a diffusion process with a Gauss distribution; the parameter $\alpha = 1$ corresponds to the Cauchy process. Thus the exponent α of the characteristic function determines the essential physical properties of the process. In particular the distributions $P_\alpha(x,t)$ do not possess finite moments of all orders. This can be seen most easily from the fact [see e.g., Montroll and West (1979)] that, for $t > 0$,

$$\lim_{|x|\to\infty} P_\alpha(x,t) \sim \frac{\alpha bt\,\Gamma(\alpha)\,\sin(\pi\alpha/2)}{\pi\,|x|^{1+\alpha}} \quad , 0 <\alpha <2 \quad . \tag{3.2.9}$$

Thus, all μ moments defined by

$$<|x|^\mu> = \int_{-\infty}^{\infty} |x|^\mu\, P_\alpha(x,t)\, dx \tag{3.2.10}$$

are finite for $\mu<\alpha$ and infinite for $\mu\geq\alpha$. In particular the variance is infinite. If the reader will recall, the existence of a finite variance was essential for the derivation of the Gauss distribution using the Central Limit Theorem. The whole notion of minimizing the quadratic deviation of a process from an assumed value depended on the average value of that quadratic function being finite. The mean square minimization which relied on local interactions in phase space and gave rise to linear algebraic relations among the parameters and moments is no longer adequate when long range interactions are present. The divergence of the variance for a Lévy process requires that a different procedures be used in the analysis of data when they are samples taken from such a process. It may occur to the reader that if the second order quantities diverge, then so too does the correlation function and its associated Fourier transform, the spectral density. It would seem therefore that such processes completely invalidate the statistical measures discussed in §2. Thus if such processes exist they must be studied differently than was suggested in §2.

Even though the Lévy distribution is characterized by apparently simple equations [cf. (3.2.6) and (3.2.7)], there are two major difficulties in understanding the physical properties of these distributions. First, the probability density $P_\alpha(x,t)$ can be evaluated in a closed form only for special choices of the parameter α. Apart from the diffusion and the Cauchy cases, a few other one-dimensional cases have been discussed by Zolotarev [see Montroll and West (1979)]. The second difficulty is that it is not possible to construct simple evolution equations of the Fokker-Planck form for the probability density, i.e., equations involving a first derivative in time and second derivative in the state variable. For a Lévy process such first derivative in time equations, in general, contain integral operators in the state variable x. For example, in one dimension when $\alpha \neq 2$, the evolution equation can be shown to be of the form [Seshadri and West (1982)]:

$$\frac{\partial P_\alpha(x,t)}{\partial t} = \frac{b\,\Gamma(\alpha+1)}{\pi}\,\sin(\pi\alpha/2) \int_{-\infty}^{\infty} \frac{dy P_\alpha(y,t)}{|x-y|^{\alpha+1}} \quad . \tag{3.2.11}$$

Thus we see that the Lévy process is non-local in the state space of the system and therefore no finite number of differentials can be used to represent the kernel $|x-y|^{-\alpha-1}$. It may be recalled from courses in electricity and magnetism or hydrodynamics that a kernel of the form $|x-y|^{-2}$ corresponds to an integral representation of a Laplacian operator, i.e., it is the

Greens function for the process. By the same token the integral operator $|x-y|^{-\alpha-1}$ is the Greens function associated with a *fractional derivative* in the state space of the system.

The evolution equation (3.2.11) can be expressed more simply by introducing the Fourier transform (3.2.7) into (3.2.11) and inverse Fourier transforming. The resulting equation for the characteristic function is

$$\frac{\partial}{\partial t}\ \phi(k,t)\ =\ -b\,|k|^{\alpha}\,\phi(k,t)\ , \tag{3.2.12}$$

whose solution is, of course, obtained by direct integration to be (3.2.6) when the initial condition $\phi(k,0)\ =\ 1$ is used.

The Lévy distribution has a number of interesting properties. Firstly it is a member of the class of infinitely divisible distributions, as was first shown by Khinchine and Lévy (1936) and Lévy (1937). This is a consequence of the fact that (3.2.7) satisfies the BSCK-chain condition, and has been shown by Jona-Lasinio (1975) to be the probability theory equivalent of the scaling condition from renormalization group theory. A second interesting property of (3.2.7) is the general scaling relation in E-dimensions,

$$P_\alpha(\beta^{1/\alpha}\,\mathbf{x},\ \beta t)\ =\ \beta^{-E/\alpha}\ P_\alpha(\mathbf{x},t) \tag{3.2.13}$$

for a real parameter β. This relation can be established directly from the definition of the characteristic function (3.2.6) and the probability density (3.2.7) for $E\ =\ 1$. It implies that if the dynamic process $X(t)$ is a random variable with probability density $P_\alpha(x,t)$ then the two stochastic functions $X(\beta t)$ and $\beta^{1/\alpha}X(t)$ have the same distribution. This scaling relation establishes that the irregularities are generated at each scale in a statistically identical manner, i.e., if the fluctuations are known in a time interval $\beta t_0 \geq t \geq t_0$, they can be determined in the expanded interval $\beta^2\, t_0 \geq t \geq \beta t_0$ as well as in the contracted interval $t_0 \geq t \geq t_0/\beta$. This scaling is known as self-similarity and is one of the defining properties of fractal processes. Note that $\alpha\ =\ 2$ corresponds to one-dimensional Brownian motion of the scalar function $X(t)$ and $P_\alpha(x,t)$ is then a zero-centered Gauss distribution with variance proportional to t.

The final property of interest is the probability that $X(t)$ is outside some interval $(-X,X)$ at fixed time t:

$$Prob\ (\,|x|>X\,)\ \sim\ \frac{constant}{X^\alpha}\quad as\quad X\rightarrow\infty\ ,\alpha<2\ . \tag{3.2.14}$$

Equation (3.2.14) indicates that the distribution for the x-process is that of a hyperbolic random variable. Such hyperbolic distributions preserve self-similarity and is shown below to have trajectories with fractal dimensionality D. Thus a Lévy process with an exponent α has a trajectory (graph of x versus t) with (Hausdorff-Besicovitch) fractal dimensionality $D = 2 - 1/\alpha$. A process described by a function, the graph of which has the Hausdorff-Besicovitch dimensionality greater than unity is said to be fractal, i.e., a fractal process is one described by a function that exhibits a fractal dimensionality [Mandelbrot (1977)]. For example, the fractal dimensionality D of a Brownian motion process is two in a Euclidian space of dimension (E) of one, two or three dimensions. In general fractal functions (processes) with $D<2$ and $D<E$ are not space filing [Mandelbrot (1977), Montroll and West (1979), Berry and Lewis (1980) and Hughes, Shlesinger and Montroll (HSM) (1981)], i.e., in order to maintain the self-similarity property a fractal process can only occupy space and/or time in clustered or localized patches.

To determine if a Lévy trajectory with homogeneous intervals is a fractal we adopt a technique described by Orey (1970) and use the "potential" definition of a fractal dimension. As described by Berry (1979) we assume a positive charge of unit density to uniformly cover the interval $(-l, l)$ along the t-axis. This charge is moved up and down in the x direction until it intercepts the trajectory $X(t)$. The "electro-static energy" of this fractal line of charge can be written by using a modified force law, ie., we use an interaction potential $\eta^{-\mu}$ rather than η^{-1} where η is the separation distance between two elements of charge on $X(t)$. This energy is $E(\mu)$ and is given by the integral

$$E(\mu) = \int_{-l/2}^{l/2} dt_1 \int_{-l/2}^{l/2} \frac{dt_2}{\left\{ (t_1 - t_2)^2 + [X(t_1) - X(t_2)]^2 \right\}^{\frac{\mu}{2}}} \tag{3.2.15}$$

and the fractal dimension D is defined as the greatest value of μ for which $E(\mu)$ is finite.

In parallel with Orey we investigate the convergence of $E(\mu)$ by restricting our analysis to "almost all" functions $X(t)$ in an ensemble of realizations by evaluating the average $<E(\mu)>$. Both Orey and later Berry use a Gaussian ensemble, whereas here we use the more general Lévy ensemble.

Denoting the separation of phase points along the trajectory by $\Delta x = X(t_1) - X(t_2)$ and the separation in time by $\Delta t = t_1 - t_2 > 0$ the probability density is

$$P_\alpha(\Delta x, \Delta t) = \int_{-\infty}^{\infty} \frac{d\kappa}{2\pi} e^{i\kappa \Delta x} e^{-\gamma |\kappa|^\alpha \Delta t} . \tag{3.2.16}$$

The average of the energy integral (3.2.15) is then written as

$$<E(\mu)> = \int\limits_{-\infty}^{\infty} d\Delta x P_\alpha(\Delta x, \Delta t) \int\limits_{-l/2}^{l/2} dt_1 \int\limits_{-l/2}^{l/2} dt_2 \frac{1}{\left[\Delta t^2 + \Delta x^2\right]^{\mu/2}}$$

$$= \int\limits_{-\infty}^{\infty} d\Delta x \int\limits_{-l}^{l} d\Delta t \frac{(l - |\Delta t|)P_\alpha(\Delta x, \Delta t)}{[\Delta t^2 + \Delta x^2]^{\mu/2}} \tag{3.2.17}$$

The important region of the integrand is $\Delta t = 0$ and $\Delta x = 0$ since this is where the convergence of (3.2.17) is determined. Since it is only the convergence of (3.2.17) that is of interest we ignore all irrelevant constants and reset the limits of the Δt integration at infinity. Thus we obtain

$$<E(\mu)> \propto \int\limits_{-\infty}^{\infty} d\Delta x \int\limits_{-\infty}^{\infty} d\Delta t \frac{P_\alpha(\Delta x, \Delta t)}{[\Delta t^2 + \Delta x^2]^{\mu/2}}$$

$$\propto \int\limits_{-\infty}^{\infty} d\Delta x \int\limits_{-\infty}^{\infty} d\Delta t \int\limits_{-\infty}^{\infty} d\kappa \frac{e^{-\gamma|\kappa|^\alpha|\Delta t|}e^{i\kappa\Delta x}}{[\Delta t^2 + \Delta x^2]^{\mu/2}} . \tag{3.2.18}$$

Integrating (3.2.18) over Δx we obtain

$$<E(\mu)> \propto \int\limits_{0}^{\infty} d\Delta t \int\limits_{-\infty}^{\infty} d\kappa \left(\frac{\kappa}{\Delta t}\right)^{\frac{\mu-1}{2}} e^{-\gamma|\kappa|^\alpha\Delta t} K_{\frac{\mu-1}{2}}(\kappa\Delta t) \tag{3.2.19}$$

where for moderate or small values of κ and Δt we use the approximation for the Bessel function $K_{\frac{\mu-1}{2}}(\kappa\Delta t) \sim (\kappa\Delta t)^{\frac{1-\mu}{2}}$ reducing (3.2.19) to

$$<E(\mu)> \propto \int\limits_{0}^{\infty} d\Delta t \int\limits_{0}^{\infty} d\kappa \frac{e^{-\gamma\kappa^\alpha\Delta t}}{\Delta t^{\frac{\mu-1}{2}}} . \tag{3.2.20}$$

Finally by transforming variables $z = \kappa^\alpha\Delta t$ we can rewrite (3.2.20) in the factored form

$$<E(\mu)> \propto \int\limits_{0}^{\infty} z^{\frac{1}{\alpha}-1} e^{-z} dz \int\limits_{0}^{\infty} \Delta t^{1-\mu-\frac{1}{\alpha}} d\Delta t . \tag{3.2.21}$$

As $z \to 0$ the first integral remains finite. As $\Delta t \to 0$ the second integral remains finite if $1-\mu\frac{-1}{\alpha} > -1$, so that the fractal dimensions of the Lévy trajectory is

$$D = 2 - \frac{1}{\alpha} . \tag{3.2.22}$$

This result agrees with the phenomenological analysis of Seshadri and West (1982). Thus for Brownian motion $\alpha = 2$, and $D = 3/2$ as is well known. Where α decreases from 2 to 1/2, D decreases monotonically from 3/2 to 0. For $\alpha < 1/2$ we choose $D = 0$.

As a final general comment we note that if $X(t)$ is a fractal process, then its graph is a discontinuous, non-differentiable function. Mandelbrot has recently drawn attention to the importance of such functions in physical as well as biological systems [Mandelbrot (1977, 1982)] and not only advances the thesis that most physical processes are discontinuous and are described by non-differentiable functions, but in addition that those processes studied by N. Wiener (1984), i.e., diffusive Gauss processes, although of this kind, are in fact benign by comparison. Thus although Wiener processes are presently encounted in practically all physical models of phenomena in which fluctuations are thought to be important (cf. §2.3 and §2.4), Mandelbrot contends that the random component of nature is much richer. The experimental data he has amassed and juxtaposed to support this contention is impressive.

3.2.2 Pareto - Lévy Tails

It is possible that in the midst of this discussion of the mathematical properties of the Lévy distribution that the connection with experimental observables is blurred. Therefore, let us re-emphasize that the dominant feature of the Lévy distribution is the existence of a long tail [cf. (3.2.9) and (3.2.14)]. Such tails are readily observed in a number of social phenomena, e.g., the social economist V. Pareto (1897) found such a distribution from his analysis of the statistics on the income and wealth of individuals from a variety of historical records. The empirical law which Pareto postulated as a result of his studies is usually given as

$$N = \frac{A}{x^\nu} \tag{3.2.23}$$

where A is a constant, ν is the Pareto coefficient, and N is the number of people having income x or larger. He found that a single value of the parameter ν (=1.5) very nearly characterized every society and every period for which data were available. Note that (3.2.23) has the same hyperbolic form as the asymptotic form of the Lévy distribution (3.2.14). Montroll and Shlesinger (1982) demonstrated that tails of the Pareto-Lévy form could be obtained exactly using a successive amplification mechanism, rather than asymptotically as we did with (3.2.14).

Let us denote a distribution having finite central moments by $g(x)$ and denote the mean value of x by $\bar{x}$. Although it is not necessary to specify the distribution $g(x)$, it has been known for a long time that throughout most of its range the distribution of income is lognormal, see e.g., Badger (1980) for a discussion. We can understand this from the argument given in §2.1 regarding the probability density for a contingency process. The probability of achieving a given level of income is contingent on a large number of factors such as training, opportunity, motivation, skill, inheritance etc. This contingency argument would successfully describe the 5-97 percentile range of the income distribution. However as Montroll and Shlesinger (1982) point out, the last 2 or 3 percentile operate in a somewhat different mode. They (the last 2 or 3 percentile) frequently collect their extreme wealth through some amplification process that is not available to the rest of us; that process varies from case to case. The argument given below demonstrates how a distribution can undergo a transition in the last few percentile of a population from a distribution with finite central moments into a universal power-law; this transition is manifest in the distribution of wealth [see e.g., Badger (1980)].

Following Montroll and Shlesinger (1982) we apply an amplification such that $g(x)$ has the new mean value $\lambda \bar{x}$, i.e.

$$g\left(\frac{x}{\bar{x}}\right) \frac{dx}{\bar{x}} \rightarrow g\left(\frac{x}{\lambda \bar{x}}\right) \frac{dx}{\lambda \bar{x}} \tag{3.2.24}$$

and assume this occurs with probability p. Applying the amplification again so that the amplification is amplified and the mean of the distribution is $\lambda^2 \bar{x}$, again with a probability proportional to p. This process is continued repeatedly resulting in the new distribution

$$G(y) = (1-p)\left\{ g(y) + \frac{p}{\lambda} g(y/\lambda) + \frac{p^2}{\lambda^2} g(y/\lambda^2) + \dots \right\} \tag{3.2.25}$$

where $y = x/\bar{x}$ and p determines the range of the underlying distribution and is chosen so that $G(y)$ is properly normalized. Note that $G(y)$ satisfies the recursion relation

$$G(y) = \frac{p}{\lambda} G(y/\lambda) + (1-p) g(y) \quad . \tag{3.2.26}$$

The complete solution of the inhomogeneous scaling equation (3.2.26) is complicated involving the use of Mellin transforms, but it is easy to obtain the desired asymptotic properties of $G(y)$. This is accomplished by examining the non-analytic part of $G(y)$ that leads to the Pareto-Lévy tail. In the case $p \rightarrow 0$ there is no amplification and $G(y)$ becomes the same as $g(y)$. If p is small, say 10^{-2}, and λ is 10 then $G(y)$ is still close to $g(y)$ since the first term in (3.2.26) may be neglected. However when y becomes large $g(y) \rightarrow 0$ since the central moments are assumed to be finite. We also assume the fall off of $g(y)$ with y is faster than that of $G(y)$. Then the asymptotic form of $G(y)$ is determined by the homogeneous scaling relation

$$G(y) = \frac{p}{\lambda} G(y/\lambda) \quad . \tag{3.2.27}$$

obtained by neglecting the $g(y)$ term in (3.2.26). We postulate a solution of the form $A(y)/y^{1+\alpha}$, which by direct substitution into (3.2.27) yields

$$\alpha = \ln(1/p)/\ln\lambda \tag{3.2.28}$$

and $A(y) = A(y/\lambda)$, i.e., $A(y)$ is an oscillatory function, periodic in log y with period log λ. Thus we obtain the probability density

$$G(y) \sim \frac{A(y)}{y^{1+\alpha}} \qquad\qquad (3.2.29)$$

and α appears here naturally as a fractal dimension [Hughes, Shlesinger and Montroll (1981); Seshadri and West (1982)]. Badger (1980), using U.S. income data from 1935/36, determined that the best value of α is 1.63 [cf. Figure 3.2.1].

If $p\lambda > 1$ then since $p > 1/\lambda$ from (3.2.29) we observe that $0 < \alpha < 1$ and $G(y)$ represents the density function of a divergent branching (bifurcating) process having an infinite mean value, i.e., the $\mu = 1$ moments in (3.2.10) diverge. If $p\lambda < 1$ but $p\lambda^2 > 1$, then $1 < \alpha < 2$ and the mean of y using $G(y)$ will be finite, the $\mu = 1$ term in (3.2.10) converges. However, the fluctuations about the mean are infinite, the $\mu = 2$ term diverges. The connection to the tail of the Lévy distribution is now clear.

3.2.3 Fractal Random Walks

In our discussion of the distribution of errors and the processing of data we maintained the underlying assumption that processes changed smoothly and that a given data point was most strongly influenced by adjacent data points. This was nowhere more clearly evident than in the simple random walk model we used to derive the Gauss distribution of errors. In that model the probability of the error increasing or decreasing at each point in time was p and q, respectively. The probability density of having a certain observed error after n time intervals is a binomial distribution which for $p=q=1/2$ in the limit $n \to \infty$ goes over to a Gauss distribution. The fundamental assumption used in the argument is that the steps are local in state (error) space, the transition probabilities up and down (p and q) are independent of the location in state space and they only connect adjacent states. These assumptions guarantee that the mean square error remains finite for finite time, a result which has pleased all modelers since de Movire (1732) first derived the distribution which now bears the name of Gauss (1809).

The style of thought of the classical physicist has been that certain general properties of physical systems are independent of the detailed microscopic dynamics. The phenomenological giant, equilibrium thermodynamics, with its first and second laws is built on this point of view. The discipline of statistical physics in large part concerns itself with understanding how microscopic dynamics, which are deterministic classically, manifest stochastic behavior in macroscopic measurements and give rise to the observed thermodynamic relations. In the past the discussions have centered on the equilibrium properties of processes, so that the traditional wisdom of equilibrium thermodynamics could be applied, and much of the microscopic dynamics could be ignored. All that was required was that: (1) the spatial dimension of the system be restricted;

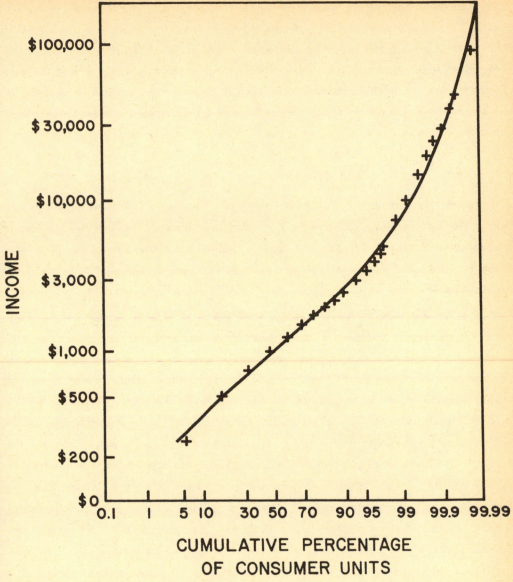

Figure 3.2.1.   Distribution of families and single individuals by income levels 1935-36 (data from U.S. Nat. Res. Com., cons. Inc. in the U.S. 1935-36, Govt. Printing Office (1939)). The values of the Pareto parameter is 1.63. [Taken from Badger (1980)].

(2) certain moments remain finite and (3) the range of the interactive force remain finite. These restrictions can be summarized by saying that there is a fundamental scale associated with the random walk model of the process. If $p(l)$ is the single-step probability density function (the "jump distribution" that replaces the p and the q in the simple nearest neighbor random walk) then in order for the qualitative behavior of the process to be insensitive to the precise form of $p(l)$ the mean-squared displacement per step

$$\sum_l l^2 p(l) = <l^2>$$
(3.2.30)

is finite. As pointed out by Hughes, Montroll and Shlesinger (1983) when this condition is violated weakly, a diffusive limit can still be exhibited, but only by introducing a nonstandard scaling of length and time to define the diffusion constant. We show below that for a sufficiently strongly divergent integral the continuum limit is the Lévy distribution.

The increased sensitivity of today's measuring instruments along with a reluctance of contempory scientists to ignore data more that two standard deviations from the mean has provided us with a picture of many complex events that are apparently without a scale of length. We see emerging patterns of structure involving clusters within clusters or waves within waves spanning many decades of traditional scale lengths. In the absence of a greatest or smallest scale length these phenomena are said to have a self-similarity property or to exhibit scaling. In §3.2.1 we discussed stochastic processes that exhibited this scaling property and related them to the fractals of Mandelbrot. HMS (1982) have provided a random walk on a lattice having a hierarchy of self-similar clusters built into the distribution function of allowed jumps. They have named these *fractal random walks* and have shown them to have the properties: (1) they give a simple realization of a fractal random process, (2) the characteristic functions have highly nonanalytic behavior at all points and (3) they lead to an analog, in a probability context, of real-space renormalization group transformation.

In the following discussion we highlight the contents of HSM (1981) which was the first paper on fractal random walks and tended to emphasize the clustering property of the walk. For a one-dimensional random walk on an infinite perfect lattice of spacing Δ, $p(l)$ is the jump distribution i.e., the probability of a step (jump) having a total displacement l, and $P_n(l)$ is the probability of occupuing the state (lattice site) l after n steps. The probability density satisfies the recursion relation

$$P_{n+1}(l) = \sum_{l' = -\infty}^{\infty} p(l - l')P_n(l') \ .$$
(3.2.31)

This equation is linear and can be solved for $P_n(l)$ for any given initial condition $P_o(l)$ by a number of standard techniques. We are not interested in those techniques here. What is of interest is the process described by $P_n(l)$ for certain choices of $p(l)$. In particular those choices of $p(l)$ for which $<l^2> = \infty$ in (3.2.30) under a symmetric random walk, but such as to give rise to a symmetric Lévy (stable) distribution in the continuum limit for $P_n(x) = \lim_{\Delta l \to 0} P_n(l)$ when $l\Delta \to x..$

One of the many possible distributions of individual step lengths $p(l)$ leading to the continuum characteristic function

$$\phi_n(k) = \int\limits_{-\infty}^{\infty} e^{ikx} P_n(x) \, dx = e^{-\gamma n |k|^\alpha} \tag{3.2.32}$$

for a Lévy process in discrete time is

$$p(x) \approx constant / |x|^{\alpha+1} \quad (0 < \alpha < 2) \tag{3.2.33}$$

as $|x| \to \infty$. When $\alpha < 2$ the second moment for this distribution is infinite. If we introduce the generating functions $\lambda(k)$, also called the structure function,

$$\lambda(k) = \sum_{l=-\infty}^{\infty} p(l)e^{ilk} \tag{3.2.34}$$

then (3.2.31) can be written as

$$\Gamma_{n+1}(k) = \lambda(k) \, \Gamma_n(k) \tag{3.2.35}$$

where

$$\Gamma_n(k) \equiv \sum_{l=-\infty}^{\infty} P_n(l) \, e^{ikl} \quad . \tag{3.2.36}$$

Iterating (3.2.35) with $\Gamma_0(k) = 1$ we obtain $\Gamma_{n+1}(k) = \lambda^n(k)$ so that the solution to (3.2.31) can be written as the inverse Fourier transform

$$P_n(l) = \frac{1}{2\pi} \int\limits_{-\pi}^{\pi} \lambda^n(k) \, e^{-ikl} \, dk \quad . \tag{3.2.37}$$

The analytic properties of the structure function $\lambda(k)$, as pointed out by Weiss and Rubin (1983) are intimately related to many interesting properties of the random walk.

To investigate the properties of $P_n(l)$ for the jump distribution (3.2.33) Weiss and Rubin (1983) study $\lambda(k)$ for small values of k. To do this they write $\lambda(k)$ as

$$\lambda(k) = \lambda_0(k) + 2A \sum_{l=1}^{\infty} \frac{\cos lk}{l^{\alpha+1}} \tag{3.2.38}$$

where A is a constant and $\lambda_0(k)$ is assumed to have a second derivative at $k = 0$, i.e., to have a Taylor expansion. For $\alpha < 2$ the second moment $<l^2>$ diverges so that the second derivative of $\lambda(k)$ does not exist at $k=0$, and for $\alpha<1$ neither does the first. They examine the behavior of the function

$$G(k) = \sum_{l=}^{\infty} \frac{1-\cos lk}{l^{\alpha+1}} \qquad (3.2.39)$$

and following the analysis of Gillis and Weiss (1970) find

$$G(k) \approx \frac{\pi k^{\alpha}}{2\Gamma(1+\alpha)\sin(\pi\alpha/2)} \qquad (3.2.40)$$

Therefore the asymptotic expression for $P_n(l)$ is

$$P_n(l) \sim \frac{1}{2\pi} \int_{-\infty}^{\infty} dk \, e^{-Bn|k|^{\alpha}} \cos k \, l \qquad (3.2.41)$$

where B is a calculable quantity. Analysis of this integral [see also Montroll and West (1979)] for $\alpha<2$ and $l \to \infty$ with n large but fixed yields

$$P_n(l) \sim \frac{\mu Bn}{\pi \, l^{\alpha+1}} \Gamma(\alpha) \sin(\pi\alpha/2) \qquad (3.2.42)$$

In Figure 3.2.2 we indicate the shape of the probability density of a Lévy (stable) process for various values of the Lévy index α.

Mandelbrot (1977) has numerically investigated a two-dimensional and three-dimensional Lévy random flight model by generating random flights with a computer. His computer simulations of such Lévy flights in two dimensions yield much more interesting trajectories in the case $\alpha<2$ than in the "diffusive" case ($\alpha = 2$ or any flight for which $<l^2>$ is finite). Several of the flights generated are displayed in Figure (3.2.3) and (3.2.4). One of the remarkable features evident in his patterns is a hierarchy of clusters. In the generation of points visited in the random flight one usually finds a cluster of points appearing. Then an occasional large displacement is experienced which begins a new cluster some distance from the old. More new clusters are generated in a similar way until a giant displacement occurs which starts a new cluster of clusters, etc. Since the dispersion in the displacement distribution is infinite (except when $\alpha = 2$) this hierarchy of clusters keeps expanding as the walk continues. Thus points visited appear in clusters, well separated in space. Under magnification, each cluster is found to consist of a set of clusters, each of which in turn is a set of clusters and so on, giving a nested hierarchy of "self-similar" clusters. Another way of picturing the sitatuion is to imagine approaching the points generated from a great distance. At first, the points will seem to be a single cluster. As one gets closer, it will be observed that the cluster is really composed of smaller clusters such that upon approaching each smaller cluster, it will seem to be composed of a set of still smaller clusters, etc. The clustering is tighter and more separated in space as α

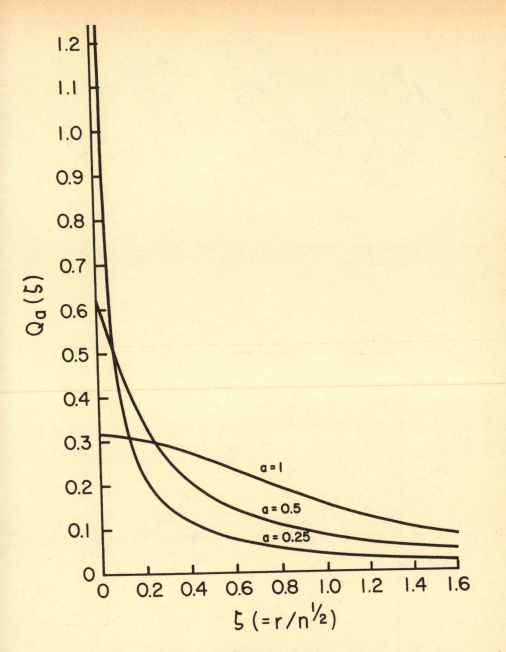

$$\xi\,(=r/n^{1/2})$$

Figure 3.2.2. Curves of $n^{1/\alpha} P_n(l) = Q_\alpha(\varsigma)$ where ς is the variable $l/n^{1/\alpha}$, for $P_n(l)$ the probability density is a stable law. The function $Q_\alpha(\varsigma)$ is

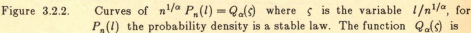

$$Q_\alpha(\varsigma) = \frac{1}{\pi} \int\limits_0^\infty e^{-v^\alpha} \cos(\varsigma v)\, dv$$

and is symmetric around $\varsigma = 0$. [See Weiss and Rubin (1983) for graph and Montroll and West (1979) for additional discussion].

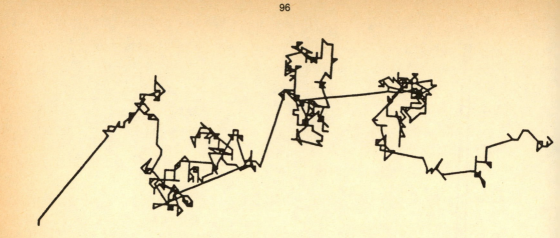

Figure 3.2.3.   Three-dimensional random flight with step lengths=determined by the Lévy distribution with index $\alpha = 1.5$. [From Mandelbrot (1977)].

Figure 3.2.4.   Cluster structure of n points of steps in a two-dimensional random walk with step length determined by the Lévy distribution with index $\alpha = 1.3$. [From Mandelbrot (1977)].

is decreased. In a Lévy flight, the trajectory spreads out over space as time increases but in a qualitatively different manner from the diffusive case.

The foregoing discussion indicates how a power-law step distribution results in a Lévy distribution for the process. HSM (1981) illustrate how self-similar clusters (Lévy distribution) can arise by examining a somewhat more intuitive step distribution. They also show how the effective dimensionality of the random walk is increased by the clustered nature of the random walk paths.

Following HSM (1981) we consider a random walk on a one-dimensional lattice of spacing Δ that exhibits clustering by constructing a discrete, but unevenly spaced, distribution of allowed step lengths:

$$p(l) = \frac{a-1}{2a} \sum_{n=0}^{\infty} \frac{1}{a^n} \left[\delta_{l,-b^n} + \delta_{l,b^n} \right] . \tag{3.2.43}$$

Here the probability of a displacement of l sites at a given step is symmetric with step lengths Δb^n (b is an integer) and probabilities corresponding to these step lengths proportional to $1/a^n$, where a, b, and Δ are positive constants and both a and b exceed unity. The structure function for this lattice walk, as given by (3.2.25), is

$$\lambda(k) = \frac{a-1}{a} \sum_{n=0}^{\infty} \frac{1}{a^n} \cos(b^n k) . \tag{3.2.44}$$

Again the nonanalytic behavior of $\lambda(k)$ as $k=0$ will determine the characteristics of $p_n(l)$. The series (3.2.44) was first considered in the last half of the nineteenth century by K. Weierstrass as an example of a function which is everywhere continuous, but nowhere differentiable with respect to k. Thus for this particular choice of the stucture function the fractal random walk was called the Weierstrauss random walk [HMS (1982)]. As mentioned by HMS (1982) Hardy (1916) has established that if $b \geq a$ then $\lambda(k)$ does not have a *finite* derivative at any value of k. A qualitative analysis of (3.2.44) based on a scaling argument will give us its dominant (singular) behavior; the more rigorous analysis is given by HSM (1981) and verifies the following result. First separate the $n=0$ term from the sum in (3.2.44) and write

$$\lambda(k) = \frac{a-1}{a} \left\{ \cos k + \sum_{n=1}^{\infty} \frac{1}{a^n} \cos(b^n k) \right\} . \tag{3.2.45}$$

We can shift the summation index n by unity in (3.2.45) to obtain the functional equation

$$\lambda(k) = \frac{a-1}{a} \cos k + \frac{1}{a} \lambda(bk) \tag{3.2.46}$$

whose singular part $\lambda_s(k)$ satisfies the equation

$$\lambda_s(k) = \frac{1}{a} \lambda_s(bk) . \tag{3.2.47}$$

We assume a solution of the form $\lambda_s(k) \propto k^\alpha$ and obtain from (3.2.47) the condition that

$$\alpha = \ln a / \ln b \quad . \tag{3.2.48}$$

The rigorous result is that the singular part of $\lambda(k)$ at $k=0$ has the form of a power k^α modulated by a oscillatory function $Q_\alpha(k)$, which is periodic in $\ln k$ with period $\ln b$. The basic idea for this argument appears to have been given by Hardy (1907), who acknowledges its suggestion to him by a Mr. J.H. Maclagan Wedderburn. The argument appears in a moden context in connection with the real-space renormalization group anlaysis of the free energy of an Ising lattice, see e.g., Schlesinger and Hughes (1981) for a comparison of the fractal random walk and renormalization group ideas.

The connection between this clustered random walk and the Lévy distribution is established by substituting (3.2.43) into (3.2.31) and taking the continuum limit. If the steps occur at equal time intervals then we can write (3.2.31) as

$$\frac{P_{n+1}(l) - P_n(l)}{\tau} = \frac{1}{\tau} \sum_{l'=-\infty}^{\infty} \left[p(l-l') - \delta_{l,l'} \right] P_n(l') \quad . \tag{3.2.49}$$

In the limit $\tau \to 0$, we obtain an equation for the probability density function at time t:

$$\frac{\partial}{\partial t} P(l,t) = \sum_{l'=-\infty}^{\infty} \lim_{\tau \to 0} \left[\frac{p(l-l') - \delta_{l,l'}}{\tau} \right] P(l',t) \quad . \tag{3.2.50}$$

The analysis is most easily performed in Fourier space:

$$\frac{\partial}{\partial t} \phi(k,t) = \lim_{\substack{\tau \to 0 \\ \Delta \to 0}} \left[\frac{\lambda(k\Delta) - 1}{\tau} \right] \phi(k,t) \tag{3.2.51}$$

where it will be recalled that $\phi(k,t)$ is the Fourier transform of the continuum probability density. It is for this reason that the limit of the lattice spacing Δ going to zero is taken in (3.2.51). HSM show that if a $\to 1+$ and b $\to 1+$ as Δ and τ to zero, such that

$$a = 1 + \mu\Delta + O(\Delta)$$

$$b = 1 + \beta\Delta + O(\Delta) \tag{3.2.52}$$

$$D_1 = \lim_{\substack{\Delta \to 0 \\ \tau \to 0}} \frac{\Delta^\mu}{\tau} = constant$$

and $0 < \mu < 2\beta$, then with D_1 a known constant (3.2.51) becomes

$$\frac{\partial}{\partial t} \phi(k,t) = -D_1 |k|^{\mu/\beta} \phi(k,t) \quad . \tag{3.2.53}$$

Equation (3.2.53) integrates to

$$\phi(k,t) = e^{-tD_1|k|^{\mu/\beta}} \tag{3.2.54}$$

so that it is the characteristic function of a Lévy distribution with index $\alpha = \mu/\beta$. Thus the fractal random walk generates a Lévy process in the continuum limit.

In the fractal random walk a hierarchy of clusters of points visited is generated, with about a subclusters per cluster and a linear scaling b between clusters of adjacent orders in the hierarchy, for a walk that is not too long. Since a and b are measures of the number of subclusters in a cluster and the spatial size of the scaling between clusters, respectively, the parameter α in (3.2.48) may be identified as the analog of a fractal dimension in the sense of Mandelbrot (1977, 1982). When the time interval over which the walk occurs increases to infinity, two outcomes of the clustering are possible; the walker may visit all the sites so that the clustering pattern is blurred out by the walker returning to fill in the spaces between clusters (persistence) or the walker may fail to visit some sites so that the clusters remain distinguishable for all time (transience). Of course this way of viewing the process is only true for lattice walks, some additional care is required when speaking of walks in the continuum. The fact that the hierarchy of clusters is not geometrically self-similar, but is self-similar in an averge sense, leads to an interpretation of $\alpha = \ln a / \ln b$ as an average fractal dimension provided the walk is transient. The detailed proof of this association between Lévy processes and fractal dimensions has been given earlier.

It is possible for the reader to test these statements regarding a process by using the computer code LEaaVY WALKS from the Appendix. It was a simple matter to modify the computer code RANDOM WALK used earlier, so that instead of yielding a Gauss distribution for the probable location of a walker, one obtains a distribution. In this code the jump probabilities are given by the power-law (3.2.14) in each of two dimensions and the user is free to specify the parameter α. Using this code one can observe how the random walk depicted in Figures (3.2.3) and (3.2.4) unfold. The dependence of the clustering property on α can be clearly seen by running the code a few times for different values of α.

3.3 Growth and Saturation

Just as linearity was found to be incompatible with the concept of a soliton, we have seen that it is also not compatible with distributions having long tails. Such distribution functions do not satisfy a Fokker-Planck equation, but instead their evolution is described by an integral equation with a nonlinear long-range kernel, i.e., $|x-y|^{-\alpha-1}$. Thus, in terms of a dynamic equation in which a system is coupled to the environment, the coupling gives rise to fluctuations that violate the assumptions made by de Moivre, Gauss and others over the past two centuries. However not all processes are modeled by stochastic equations or are described by distribution functions. Just as Newton's modeling of sound waves in the air was in terms of a set of deterministic equations, so too many other phenomena in natural philosophy are modeled by equations without fluctuations. It is often the case that these systems, in which the fluctuations are ignored, are not linear. It is important to note, based on our earlier discussion about the inability to isolate a system, that all systems in natural philosophy are fundamentally stochastic. Thus special circumstances are required in order for a "deterministic" description to be adequate. To underscore this we will present a brief account of a particular process that has some significance to nearly everyone and contrast how different mathematical modes of thought have coincided with different interpretations of the process. The process is that of growth and saturation.

3.3.1 Population and Prediction

From the beginning of the Christian era, the size of the human population grew gradually for about sixteen centuries and then with an increasing rate through the nineteenth century. Records over the period 1200-1700, while scanty, indicate some fluctuations up and down, but no significant trend is observed. By the end of the eighteenth century it had been observed by a number of people that the population of Europe seemed to be doubling at regular time intervals, a phenomenon characteristic of exponential growth. The 1798 work of Thomas Robert Malthus is most often quoted in this regard; however, Thomas Jefferson and others had made similar observations. It is curious that this first work on population was in fact written by an economist writing a discourse of moral philosophy and who had not made an original contribution to the theory of populations. The contribution of Malthus was in fact the exploration of the consequences of the fact that a geometrically growing population will always outstrip a linearly growing food supply, resulting in overcrowding and misery. Of course there was no experimental evidence for a linearly growing food supply, this was merely a convenient assumption.

Boulding (1959) refers to Malthus' view of the result of unchecked growth as the "Utterly Dismal Theorem." Only half in jest, Boulding formulates this theorem as follows: any technical improvement can only relieve misery for a while, for as long as misery is the only check on population the improvement will enable population to grow, and will merely enable more people to live in misery than before. Thus Malthus maintained that in the end all improvements only lead to an increase in the equilibrium population, which is an increase in the sum total of human misery. From his perspective, the human condition was tragic and apparently unchangeable. But even so he states, "Evil exists in the world, not to create despair, but activity." The activity we contemplate in the present work is undoubtedly something that was outside the experience of Malthus.

Malthus proposed that the incremental increase in the size of the human population was geometric, that is to say that the human species would increase in the ratio of 1, 2, 4, 8, 16, 32, etc. between successive generations. Today we understand that such growth can be represented by the elementary differential equation

$$dN(t)/dt = kN(t) \qquad (3.3.1)$$

where $N(t)$ is the population at time t and k is a constant that characterizes the rate at which the population grows. The solution to Malthus' growth equation is

$$N(t) = N(0)\exp(kt) \qquad (3.3.2)$$

where $N(0)$ is the size of the population at time $t=0$. It is clear that the solution (3.3.2) depicts a population level that grows without bound as time goes to infinity as we show in Figure (3.3.1). It was of course this unrestricted growth that prompted Malthus and others to decry the human condition as inevitably resulting in the misery produced by overpopulation.

The King of Holland was alarmed by the essay of Malthus and fearing that his small kingdom was in peril he commissioned a re-examination of the overpopulation question. The scientist Verhulst responded to the King's request and put forth a theory that somewhat mediated the pessimistic view of Malthus. In 1844 he published a work noting that the growth in population was not unbounded as modeled by (3.3.1). He argued that such factors as the availability of food, shelter, sanitary conditions, etc. all restrict (or at least influence) the growth of the population. Verhulst therefore modified (3.3.1) to provide a mechanism for the retardation of population growth. The equation he constructed is

$$dN(t)/dt = kN(t)[1 - N(t)/M] \qquad (3.3.3)$$

which limits the growth of the population to the level M. The solution to the Verhulst equation is

$$N(t) = MN(0)/\left\{ N(0) + [M - N(0)]\exp(-kt) \right\} \qquad (3.3.4)$$

which has the sigmoidal shape shown in Figure (3.3.2). Note that the Verhulst solution approaches that of Malthus as M becomes infinite, i.e., when the upper bound of the population becomes very large. Alternatively one can see that for low population levels $N(t) \ll M$ the quadratic (nonlinear) term in (3.3.3) can be neglected so that at early times the sigmoidal solution is well represented by an exponential curve.

Because the two solutions coincide at early times, that is to say at population levels far below the asymptotic value of (3.3.4), it is not difficult to fit limited data sets to

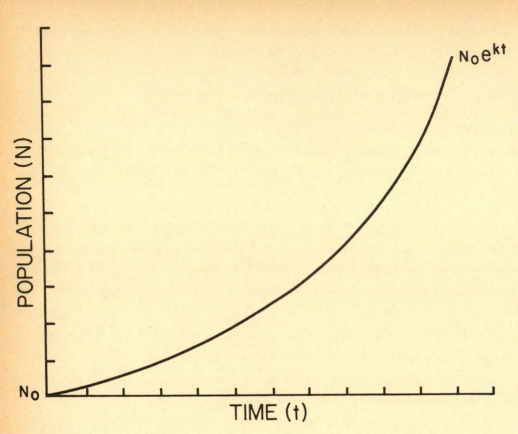

Figure 3.3.1.   A typical exponential (Malthusian) growth of population with time is
depicted.

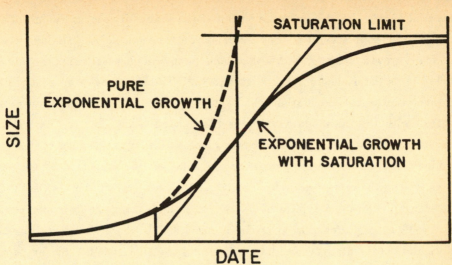

Figure 3.3.2. The general form of the logistic (Verhulst) curve is shown and compared with exponential (Malthus) growth [from D.J. de Sola Price
(1963)].

exponential curves. By this we mean that if the available data is on the initial growth region for the population and one assumes an exponential growth law, then the unbounded growth feared by Malthus and many of his contemporaries will be predicted [note the dashed curve in Figure (3.3.2)]. This does not mean that the subsequent evolution of the population will be exponential, but only that such growth is consistent with the initial data. On the other hand if one is more optimistic and assumes that a saturated growth will take place then this initial data can be used to determine the eventual largest population M that can be supported by the environment as well as the rate of growth. The larger the data set, the better the prediction of the saturation level. A typical fit to population data using the logistic curve (3.3.4) made by Pearl and Pearl is given in Figure (3.3.3) where the three parameters $M, N(0)$ and k are fit by three data points and the subsequent U.S. census data is predicted by the resulting logistic curve.

It is interesting to note the deviation in the census data from the prediction of the logistic curve. One can of course interpret this deviation as a limitation of the validity of the model, which is true. An alternative might be to view the dynamic equation (3.3.3) as a fundamental law of population growth in the absence of any biological "force." The deviation of the growth curve from the sigmoidal form then becomes evidence for the existence of such a "force" acting on the population as suggested by de Sola Price (1963). In the present case the deviation of the data from the logistic curve occurs at the time of the Great Depression in the United States when the economic conditions strongly influenced the birth rate and thus the growth rate of the population is observed to be lower than predicted. The next significant deviation arises after World War II with the return of the soldiers from the military; the increase in the birth rate more than compensated for the decline during the depression. This deviation from the predicted behavior clearly indicates a sensitivity of the growth curve to sociological and economic conditions. Montroll (1978) has proposed a law of sociology in which this sensitivity is interpreted as a "sociological force." We will have occasion to discuss these and other related ideas subsequently.

The sudden steep rise of population in the twentieth century is in part a consequence of the scientific-technologic-industrial revolution. This revolution has had the effect of making it possible to sustain a human population far larger than ever before in history. Part of this growth can be traced to advances in agricultural techniques, also to the development of suitable delivery systems both to and from the farms as well as the invention of food preservation methods usable in urban settings. This relatively recent sharp increase in the size and distribution of human populations has

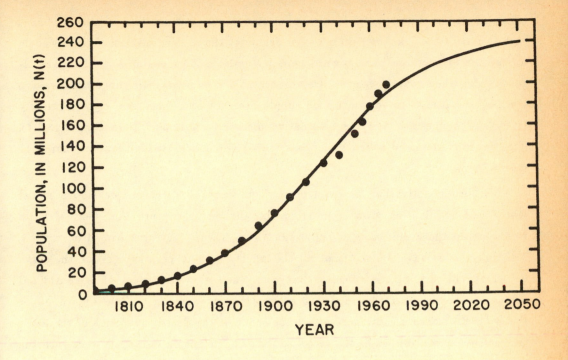

Figure 3.3.3. Population of the United States fitted by a logistic curve $N(t) = 246.5 \times 10^6 / \{1 + 2.243 \exp[-0.02984(t-1900)]\}$ so that the observed points at 1540, 1900 and 1960 are exact [see Montroll and Badger (1974)]. Points represent census data.

precipitated the question, "Will the population curve continue to rise at its present rate (Malthus), will it bend over into a sigmoidal shape (Verhulst) or will it become unstable and decrease back toward zero population?" An example of the latter is goats, who when natural enemies are absent, often overgraze their range thus leaving themselves without an adequate food supply over the long term [see, e.g., Kormandy (1969)]. The question and the alternative answers are implicitly in Figure (3.3.4). This figure is discussed at length by Salk (1983) in a much broader context than considered here.

Verhulst's motivation for constructing equation (3.3.3) was to develop a model which had self-limiting growth, in order that he might capture the effect of finite resources on the evolution of a population. The growth curve resulting from such a logistic equation has a symmetry about the inflection point, i.e., the point on the curve at which the sign of the tangent changes. Such a symmetry may be considered suspect in the quantitative modeling of growth processes. In a little known paper Feller (1940) makes a cautionary plea to not confuse the descriptive value of the logistic equation, since it does fit the recorded data of population growth so well, with the correctness of the underlying biological assumptions. He contends, and rightly so, that:

> "... the recorded agreement between the logistic and actually observed phenomena of growth does not produce any significant new evidence in support of the logistic, beyond the great plausibility of its deduction. Nor is the closeness of the agreement as strange as it appears at first sight since equally good results can be obtained by applying other mathematical forms, corresponding to quite different biological assumptions."

Feller goes on to point out that this does not impune the theory of growth, but does decry extravagant claims made with regard to the biological nature of growth and the factors which control it.

A more general expression than the logistic equation may be introduced which models the growth of an isolated species in an environment with limited resources by the rate equation

$$dN(t)/dt = kN(t)G(N/M) \quad . \tag{3.3.5}$$

Here $G(x)$ is a saturation inducing function such that

$$G(x) \rightarrow 0 \quad \text{as} \quad x \rightarrow 1 \tag{3.3.6}$$

and $G(x)$ is interpreted as the growth law for the species. The Verhulst or logistic equation may be classified as a limiting form of a particular selection for the general

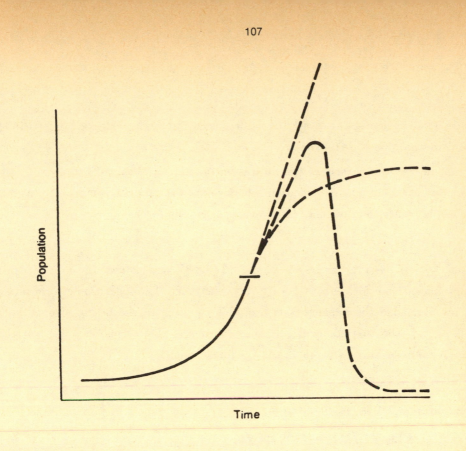

Figure 3.3.4.   The three conceptual alternatives for the growth of a population
beyond the inflection point are indicated by the dashed lines.

growth law:

$$G(x) = -(x^\alpha - 1)/\alpha \qquad (3.3.7)$$

with $\alpha = 1$. This general growth law was first used in the excellent review article by Goel, Maitra and Montroll (1971).

A second traditional form of the saturation function is obtained for $\alpha = 0$ and is due to Gompertz (1825). His equation was developed to model the mortality rates in elderly people rather than population growth and has the form,

$$dN/dt = -kN \ln(N/M) \qquad (3.3.8)$$

since $\lim(x^\alpha - 1)/\alpha \rightarrow \ln x$ as $\alpha \rightarrow 0$. Equation (3.3.8) determines the rate of death as a function of time for persons born in a given year. The logarithmic form of the "force law" impeded the implementation of this equation in the fitting of population data prior to the development of the computer. In terms of the variable $V = ln(N/M)$ we can rewrite (3.3.8) as the linear equation

$$dV/dt = -kV \quad . \qquad (3.3.9)$$

The solution to (3.3.9) is clearly given by

$$V(t) = \ln(N(t)/M) = V(0)\exp(-kt) \qquad (3.3.10)$$

or in terms of the original population variable

$$N(t) = M\exp\left\{\ln[N(0)/M]\exp(-kt)\right\} \qquad (3.3.11)$$

The solution (3.3.11) is plotted in Figure (3.3.5) for an arbitrary choice of parameter values. We see from this figure that the solution to the Gompertz equation also has the sigmoidal form. Here we further note the quantitative changes in the growth curves predicted by (3.3.5) with growth laws of the form (3.3.7) for other various values of the parameter α. In Figure (3.3.5) we depict these growth curves and emphasize that they all share the sigmoidal shape, but have slightly different symmetry properties.

A substantial number of other applications have been made of the logistic equation (3.3.3). Denoting the fraction of the asymptotic level M attained in the time t by $X(t) = N(t)/M$ we rewrite (3.3.3) as

$$\frac{dX(t)}{dt} = kX(t)[1-X(t)] \quad . \qquad (3.3.12)$$

The solution to (3.3.12) can be written as

$$\log\left[\frac{X}{1-X}\right] = \log\left[\frac{x_0}{1 = x_0}\right] + kt \qquad (3.3.13)$$

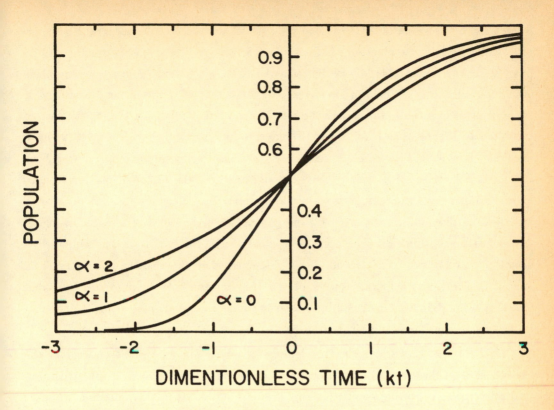

Figure 3.3.5.   The solution the the growth Equation (3.3.5) as a function of the dimensionless time (kt) is depicted for three values of the parameter α.

where x_0 is the value of $X(t)$ at time $t=0$. Note that on logarithmic graph paper a plot of $X/(1-X)$ as a function of t becomes a straight line. Fisher and Pry (1971) have made the remarkable observation that when a new industrial process or product is introduced and takes hold of the market to the extent of representing 10 percent of the output, it eventually takes over the rest of the market following the logistic law as exhibited in Figure (3.3.6) for several replacements. Montroll and West (1972) have discussed how such a process can arise using a biased random walk model which we discuss in a later section. Sahal (1981) investigates the general conditions for technological change by introducing two alternative pictures. The first has to do with a system-wide disequilibrium caused by the gap between the old and new technologies. The logistic equation describes how the system closes the gap from one equilibrium situation to the other. The second model of technological substitution concentrates on *spatial* rather than temporal aspects of the phenomenon. As pointed out by Sahal (1981) it was originally developed by biologists in the studies of morphological changes in organisms [see Reeve and Huxley (1945)].

Each of the above biological models was constructed to predict or explain the growth of a single species in isolation (including human) in an environment with limited resources. A growing population can circumvent the saturation inherent in (3.3.7) by redistributing (diffusing) into nearby unoccupied territory as the resources are depleted at their existing location. This can be observed in all processes from the spreading of bacteria in a petri dish to the settlement of pioneers in the western United States during the Nineteenth century. The diffusion does not stop the saturation, but it does slow it down in a given region of space. This effect is considered briefly here and again in §4.2.

In these examples the population becomes a function of the two dimensional vector $\mathbf{x} = (x,y)$ locating the population on the plane, i.e., $N(t) \rightarrow N(\mathbf{x},t)$. If we now redefine M as the maximum population that can occur per unit area, and if the region of space is homogeneous in an area A, then the population level density can be defined as

$$\rho(\mathbf{x},t) = N(\mathbf{x},t)/MA \tag{3.3.14}$$

and the equation of evolution including diffusion can be written

$$\partial_t \rho(\mathbf{x},t) = k\rho(\mathbf{x},t) \, G\Big(\rho(\mathbf{x},t)\Big) + D \, [\, \partial_{xx} + \partial_{yy}] \, \rho(\mathbf{x},t) \tag{3.3.15}$$

as a direct generalization of (3.3.5). Let us restrict our considerations to the one-dimensional form of (3.3.15), i.e., $\mathbf{x} \rightarrow x$. Suppressing the arguments of $\rho(x,t)$ for the moment we chose as a saturation function

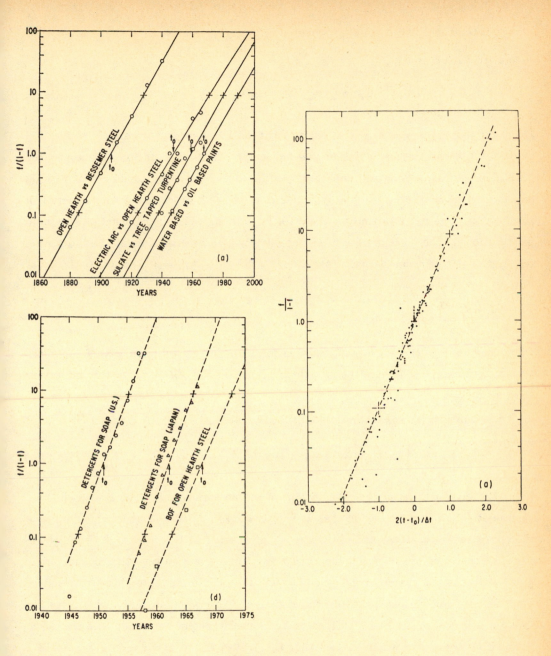

Figure 3.3.6.   Substitution data and fit to model for a number of products and processes in (a). Fit of substitutional model function to substitution data for all 17 cases vs. normalized units of time in (b): [Fisher and Pry (1971)].

$$\rho G(\rho) = (2\rho - 1) \, \partial_x \rho \tag{3.3.16}$$

so that (3.3.15) becomes

$$\partial_t \rho = \partial_x \left\{ -k\rho(1-\rho) + D \, \partial_x \rho \right\} . \tag{3.3.17}$$

Equation (3.3.17) is not a new one. It is the basic equation for cascade separation processes such as thermal diffusion and molecular effusion separation of isotopes, see Majumdar (1951) or Montroll and West (1979). It is interesting to note that (3.3.17) can be derived as the continuum limit of a random walk processes in which there is a bias, i.e., a preferred direction for the walkers to step, and a dependence of the jump distribution on the local density of the population [see §4.2.2]. One of the interesting features of this equation is that in the steady state with $\partial_t \rho = 0$ we recover the equation

$$\frac{d\rho}{dx} = \frac{k}{D} \, \rho(1 - \rho) \tag{3.3.18}$$

which has exactly the same mathematical form as the Verhulst equation for population growth except that here the growth is a function of position along the line rather than a function of time.

A possible social application of (3.3.18) and its solution, where $\rho_0 \equiv \rho(x=0)$,

$$\log \left[\frac{\rho}{1-\rho} \right] = \log \left[\frac{\rho_0}{1-\rho_0} \right] + kx/D \tag{3.3.19}$$

is for the investigation of the development of ethnic attitudes in American cities. On certain streets in large cities one has at various times had situations in which one ethnic group has lived at one end of the street and another at the other end. The "thickness" of the interface in which the change from one group to the other occurs would give a measure of the ratio k/D in (3.3.19). One could compare these values from year to year for various pairs of ethnic groups. The quantity D/k which has the units of a length might be called the penetration distance of one group into the other. A small (D/k) value would suggest that the groups try to avoid each other, while a large value would indicate that they are quite compatible. Common names associated with various ethnic groups could be used as tracer elements in examining all residential patterns through old city directories. Current appraisal could be made through detailed census of residents on various sheets.

As Montroll and West (1972) point out, another nice feature of the population equation (3.3.17) is that it can be solved exactly in the non-steady-state by applying a nonlinear transformation in the dependent variable. Suppose that $u(x,t)$ is a solution of the simple diffusion equation (2.4.12), i.e.,

$$\partial_t u(x,t) = D \, \partial_{xx} u(x,t) \ . \tag{3.3.20}$$

Then we relate the solution $u(x,t)$ to $\rho(x,t)$ through the nonlinear transformation

$$u(x,t) = \exp\left\{ \int^x g\big(\rho(x',t)\big) \, dx' \right\} \tag{3.3.21}$$

where $g(\rho)$ is an arbitrary, as yet undefined, function of the population density ρ. Then one finds with $g'(\rho) \equiv \partial g / \partial \rho$ that

$$\int^x g'(\rho)\partial_t \rho \ dx' = D \left\{ g'(\rho) \, \partial_x \rho + g^2(\rho) \right\} \tag{3.3.22}$$

where, after differentiation with respect to x, we obtain

$$\partial_t \rho = D \left[\partial_{xx} \rho + \frac{g''}{g'} (\partial_x \rho)^2 \right] + 2 D g \, \partial_x \rho \ . \tag{3.3.23}$$

Since the special case

$$g(\rho) = \frac{k}{D} (\rho - \frac{1}{2}) \tag{3.3.24}$$

yields the evolution equation (3.3.17), we find the solution to (3.3.17) is obtained from (3.3.21) after noting that

$$\partial_x \ln u = \frac{k}{D} \left[\rho - \frac{1}{2} \right] \tag{3.3.25}$$

and at time $t=0$

$$u(x,0) = \exp\left\{ \int^x \frac{k}{D} \left[\rho(x',0) - \frac{1}{2} \right] dx' \right\} \ . \tag{3.3.26}$$

Hence the solution to (3.3.17) is obtained from (3.3.25) to be

$$\rho(x,t) = \frac{1}{2} + \frac{D}{k} \, \frac{1}{u(x,t)} \, \partial_x \, u(x,t) \tag{3.3.27}$$

where $u(x,t)$ is the solution to the simple diffusion equation (3.3.20) with the initial condition given by (3.3.26). Montroll and Newell (1952) have investigated a number of choices for $g(\rho)$ in detail.

3.3.2 Scaling and Saturation

Up to this point we have considered growth and saturation in a rather limited context. However the idea that all growth processes are intrinsically limited is a fairly universal one and applies to all aggregation processes from the growth of stars down to the formation of molecular chains. In keeping with our middle of the road policy we shall avoid these two extremes and concentrate primarily on the limitations to the growth of biological organisms. A brief discussion of the reason why plants and animals are of a certain size will enable us to re-introduce the important concept of scaling, but now in a deterministic rather than a stochastic context. This idea has been used with great success by scientists such as D'Arcy Thompson (1915) to disclose the connection between biology and physics at its most rudimentary level.

In the theory of measurement it is well established that if x is an independent variable (measurement), y a derived variable and G a continuous *unknown* relationship between them, i.e., $y=G(x)$, then if both x and y are ratio scales then G has the form of a *power-law* function. For example the phenomenological relations in physics given by Ohm's law, Dalton's principle, Newton's law of gravitation, etc. can each be *formally* derived from this principle of measurement theory, see for example the discussion by Roberts (1979). As an explicit example of a physical system with ratio scales let us consider the relationship between the areas of two geometrical objects. If l_1 is the linear dimension of object 1, and l_2 the linear dimension of object 2, then the ratio of areas is given by

$$A_1/A_2 = (l_1/l_2)^2 \qquad (3.3.28)$$

where only the ratios A_1/A_2 and l_1/l_2 have a concrete physical meaning. The symbols in (3.3.28) represent the physical quantities whereas when we write $A_1/A_2 = l_1^2/l_2^2$ we interpret the equation as a relationship between the numbers associated with the physical quantities. We therefore usually abbreviate (3.3.28) by expressing the area as

$$A \sim l^2 \qquad (3.3.29)$$

where the proportionality constant depends on the system of units selected. A similar argument can be constructed for the volume (V) of an object in a Euclidean space of three dimensions,

$$V \sim l^3 \quad . \qquad (3.3.30)$$

These arguments all assume a uniform distribution of mass within the real objects of physical interest. This is in fact an assumption which must be tested in each case.

We have space here to review only a few of the amazing consequences of the simple relations (3.3.28) and (3.3.30) when applied to a biological system. Let us consider the process of growth as an increase in length, then although the increase in volume is part of the same process, because weight is proportional to volume, the effect of increasing volume is much more dramatic than increasing length. Therefore as stated by Thompson (1915), knowing the specific gravity of a species, i.e., knowing the proportionality constant β between weight W and volume V, $W = \beta l^3$ allows one to "weigh an organism with a measuring rod." The philosophy of Newtonian physics imposes no limitations of scale on our concepts, so that the application of physical laws and the deductions therefrom should be valid from one extreme of magnitude to the other. Thompson points out that there are, however, intrinsic scales for things having to do with their relation to the whole environment. "Everywhere Nature works true to scale and everything has its proper size accordingly." In particular, different physical forces act on different aspects of an object. For example some forces act directly on the surface of a body as does wind resistance, while others act directly on the mass of the body, as does gravity. The former force is proportional to area whereas the latter is proportional to volume, and it is the balance between the environmental forces produced by the different dependences on the linear dimension l that determine the optimum scale of the object. This is known as the "Principle of Similitude" or of dynamic similarity.

As concrete examples we know that the strength of a muscle varies with its cross-sectional area as does the resistance of a bone to a crushing stress. An animal in a gravitational field experiences a force in direct proportion to its volume (mass). Thus the balance between the resistance to gravity which varies as l^2 and the force of gravity which varies as l^3 sets a definite limit on the linear dimension l. Consider the more fortunate situation of a fish, whose movements are not opposed by gravity but by "skin friction." Because the velocity a fish attains (v) depends on the work (W) that the fish does to overcome the resistance (R) of the water we can write

$$W \sim Rv^2 \ . \tag{3.3.31}$$

Now the work the fish can do is proportional to its mass $(W \sim l^3)$ and the water resistance is proportioned to its surface area $(R \sim l^2)$, so that we obtain from (3.3.31)

$$v^2 \sim \frac{l^3}{l^2} = l \quad ; \quad v \sim \sqrt{l} \ . \tag{3.3.32}$$

Thus the maximum velocity the fish can attain is proportional to the square root of the linear dimension. This is *Froude's Law* for the speed of ships (or fish).

Such "intuitively obvious" scaling relations are not always observed experimentally however. For example, consider a three (Euclidean) dimensional distribution of mass points such that the quantity of mass $M(l)$ contained in a sphere of radius l increases with distance as

$$M(l) \sim l^F \quad , \quad F < 3 \ . \tag{3.3.33}$$

The case $F = 3$ is the familiar situation for a uniform distribution of mass points. However a self-similar distribution of mass points is described by a value of F less than three. In the usual case, when the mass is within a sphere of radius l and we have no information on scales below l, then the mass is assumed to have a uniform distribution throughout the volume. If now we examine the sphere on a finer scale, $l' = l/a$ say, we discover that what we had considered to be a single sphere actually consists of b smaller spheres each of radius l/a. If this process of increasing resolution is continued indefinitely we arrive at the functional relation between the mass at adjacent scales l and l':

$$M(l) = bM(l') = bM(l/a) \tag{3.3.34}$$

If we chose the functional form (3.3.33) for the mass then we find

$$F = \ln b / \ln a \tag{3.3.35}$$

as a solution to (3.3.34). As before, the quantity F is called the fractal dimensionality. Equation (3.3.33) does in fact describe the observed distribution of galaxies in the universe [see e.g., Peebles (1980)] and was first derived, using a random walk argument, by Mandelbrot (1975).

A more pedestrian example of scaling is provided by the branching of the arterial system. Cohen (1954, 1955) considered how the branches should be arranged in space in order to insure an adequate supply of blood to every element of the tissue. Rashevsky (1960) refined the arguments of Cohen using an "equivalent bifurcation system." An associated question, and the one considered here, concerns the reduction in size of the branches in the arterial system between successive bifurcations.

To determine the ratio of radii between successive generations in the arterial system we calculate the flow resistance in an idealized system of bifurcating tubes. Following Rashevsky (1962) we note that if Δp is the pressure drop along a branch, and l is the length of the branch, r its radius, $\bar{v}$ the average flow velocity of the blood, η the viscosity and ρ the density of the blood, then assuming that Poiseuille's law for pipe flow is appropriate for an artery, we have

$$\Delta p = \frac{8\eta l \bar{v} \rho}{r^2} . \tag{3.3.36}$$

Then if Q is the total flow, the measure of the resistance to the flow Ω is given by

$$\Omega = \frac{\Delta p}{Q} . \tag{3.3.37}$$

Since the total flow is the cross sectional area (πr^2) times the density ρ and the average flow velocity $\bar{v}$; $Q = \pi r^2 \rho \bar{v}$, we can write the pressure difference in terms of Q

$$\Delta p = \frac{8\eta l Q}{\pi r^4} \tag{3.3.38}$$

and the resistance to the flow in a tube of length l and radius r as

$$\Omega = \frac{8\eta l}{\pi r^4} \quad . \tag{3.3.39}$$

The total resistance of N identical tubes in parallel is $1/N$ of the resistance of a single tube. If Ω denotes the resistance of the aorta, Ω_1 that of the first branch, ..., Ω_N that of the N th branch, then since there are 2^N branches in the N th "generation" because each generation doubles the number of branches, the total resistance of the system is

$$\Omega_T = \Omega_0 + \frac{1}{2}\Omega_1 + \dots \frac{1}{2^N}\Omega_N \quad . \tag{3.3.40}$$

In each of these contributions we can substitute the relation from (3.3.39). We assume that all branches in the j th generation have the same length l_j and that the wall of this branch is proportional to the radius r_j, i.e., $\Delta_j = \alpha r_j$, so that setting the density of the wall to unity the total mass of the system after N bifurcations is

$$M = 2\pi\alpha \, r_0^2 \, l_0 + 2\pi\alpha \sum_{j=1}^{N} 2^j \, r_j^2 \, l_j \quad . \tag{3.3.41}$$

Solving (3.3.41) for r_0 and (3.3.39) for Ω_j and substituting those expressions into (3.3.40), we obtain the total resistance

$$\Omega_T = \frac{8\eta}{\pi} \frac{l_0^2}{\left(M/2\pi\alpha - \sum\limits_{j=1}^{N} 2^j r_j^2 \, l_j\right)^2} + \frac{8\eta}{\pi} \sum_{j=1}^{N} \frac{1}{2^j} \frac{l_j}{r_j^4} \quad . \tag{3.3.42}$$

We now apply a principle formulated by Rashevsky: *For a set of prescribed biological functions an organism has the optimal possible design with respect to economy of material used, and energy expenditure, needed for the performance of the prescribed functions.* This principle is implemented through a minimization argument on the total resistance just as it was used to determine the "law of errors" and in the method of least squares in §2. Setting the partial derivative of Ω_T with respect to r_p, $p = 1, 2, \cdots N$, equal to zero we arrive at the relation

$$r_j = r_0/2^{j/3} \tag{3.3.43}$$

for the relative radius of the j th generation of the arterial systems. The ratio of the radii of successive generations is

$$\frac{r_{j+1}}{r_j} = \frac{1}{2^{1/3}} = 0.794 \quad . \tag{3.3.44}$$

This compares favorably with the observed ratio of 0.8. Notice that we have shifted the emphasis in these final examples from that of a modifiable saturation to that of a guiding principle governing the overall development of a complex process or system to some fundamentally optimal size in an earth environment.

3.4 The Relaxation Oscillator

Nature abounds with rhythmic behavior that closely intertwines the physical, biological and social sciences. The spinning earth gives rise to periods of dark and light that are apparently manifest through the circadian rhythms in biology. An incomplete list of such daily rhythms is [see e.g., Luce (1971)]: the apparent frequency in fetal activity, variations in body and skin temperature, the relative number of red and white cells in the blood along with the rate at which blood will coagulate, the production and breakdown of ATP (adenosine triphosphate), cell division in various organs, insulin secretion in the pancreas, susceptibility to bacteria and infection, allergies, pain tolerance, and on and on. No attempt has been made here to distinguish between cause and effect, since this would lead us into the area of physiology (from the examples given so far). Here we only stress the observed periodicity in each of these phenomena. The shorter periods associated with the beating of the heart and breathing are also modulated by a circadian rhythm.

These periodicities ranging from seconds to days only represent a narrow region of the spectrum of frequencies manifest in biological processes. For example many illnesses have a rhythmic component that is monthly rather than daily, mood cycles in men, premenstrual syndrome, periodic hypertension, etc. As summarized by Luce:

"Although modern society presumes constancy of behavior, this may, in fact not exist in most human beings. A true base line for mood and behavior might reveal cycles of several frequencies: near-monthly, seasonal, and annual."

This observation may have strong implications in economics, sociology and the other sciences concerned with the interaction of humans in large numbers. The implications in the areas of medicine and mental health are being systematically investigated by an increasing number of scientists.

There is a tendency to think of the rhythmic nature of many biological phenomena, such as the beating of the heart, breathing, circadian rhythms, etc. as arising from the dominance of one element of a biosystem over all the other elements. A logical consequence of this mode of thought is the point of view that much of the biosystem is passive, taking information from the dominant element and merely passing it along through the system to the point of utilization. This perspective is being called into question more and more by the mathematical biologists, a substantial number of which regard the rhythmic nature of biological processes to be the consequence of a dynamic interactive nonlinear network, that is to say that biological systems are systemic. The mathematical models used to support this contention were first developed in the evolving discipline of *nonlinear dynamics* which appears to be emerging as a new branch of theoretical physics and applied mathematics, [see e.g., the review edited by Jorna (1978) or the text of Lichtenberg and Lieberman (1983)]. The application of some of the techniques of nonlinear dynamics to biological oscillations is of recent origin being championed by Winfree (1977, 1984), Glass et al. (1982, 1983), West et al. (1985) among many others. However the application of nonlinear equations to describe biorhythms dates back to the 1929 work of van der Pol and van der Mark on the *relaxation oscillator.*

In discussions of this kind it is always important to be cognizant of the role of mathematical modeling in the development of an understanding of the basic phenomenology. The theory of evolution provides an example of a developing body of knowledge that was guided, at least in part, by the mathematical ideas applied in the study of genetics. The true influence of mathematics on the theory of evolution was unclear even to those making the major contributions. Provine (1977) points out that these mathematical models of population genetics had a significant influence upon evolutionary thinking in at least four ways:

"First these models demonstrated to most biologists in the 1930's and 40's that Mendelism and natural selection, plus processes known or reasonably supposed to exist in natural populations, were sufficient to account for micro-evolution at the population level. Second, the models indicated that some paths taken by evolutionary biologists were not fruitful. Third, the models elucidated, complimented, and lent greater significance to the results of field researchers including the work of systematists and provided the intellectual framework for later field research. And finally I conclude that the influence of the models is greater than many biologists think, because their impact was sometimes invisible to those influenced."

Although Provine goes to great lengths to support these remarks in the evolutionary theory context, I contend that these remarks remain valid throughout natural philosophy. In particular as Provine emphasizes

" . . . mathematical geneticists and their models stimulated and guided field researchers on natural population by identifying clearly some relevant parameters of the evolutionary process, and by elucidating possible structures and dynamics of natural populations, thus influencing experimental design and generating, directly or indirectly, a large number of hypotheses testable in the field."

Let us now return to the rhythmic behavior of man and his environment discussed above. In addition to naturally occurring rhythms such phenomena as periodic catatonia, periodic hypertension, manic depression cycles, and so on indicate that certain periodicities denote pathologies. Luce (1971) comments:

" The concept of an internal oscillation as a pathological state, replacing the normal state of metabolic activity has many implications for treatment. By understanding the temporal aspects of the illness, it may become possible to schedule medication so as to restore normal rhythmicity in the metabolic system."

Oscillations in biological processes do not in general follow a simple harmonic variation in space and/or time. The more usual situation is one in which the period of oscillation is dependent on a number of factors, some intrinsic to the system but others external to it, such as the amplitude of the oscillation, the period at which the biological unit may be externally driven, the internal dissipative properties, fluctuations, and so on. In particular since *all* biological systems are thermodynamically open to the environment they are dissipative, i.e., they give up energy to their surroundings in the form of heat. This regulatory mechanism helps to maintain the unit at an even temperature. Thus, if a simple harmonic oscillator is used to realistically model a biological unit undergoing oscillations, it must contain dissipation. It is well known however, that the asymptotic trajectory of a dissipative harmonic oscillator is a stable fixed point in phase space. This means that the amplitude of the oscillator excursions become smaller and smaller until eventually it comes to rest. For the oscillator to remain periodic, energy must be supplied to the system in such a way as to balance the continual loss of energy due to dissipation. If such a balance is maintained, then the phase space orbit will become a stable *limit cycle*, i.e., all orbits in the neighborhood of this orbit will merge with it asymptotically. However such simple oscillators do not have the appropriate qualitative features for describing biological systems. One of the important properties that such a linear oscillator lacks and which is apparently ubiquitous among biological units is that of being self-starting. Left to itself a bio-oscillator will spontaneously begin to oscillate apparently without external excitation. One observes that the self generating or self-regulating characteristic of bio-oscillators depends on the intrinsic nonlinearity of the biological unit. In summary we can say that biological oscillations are nonlinear and the nonlinearities arise in a number of interesting ways.

3.4.1 Glossary of Nonlinear Terms

It is useful to have a glossary of frequently encountered nonlinear terms; not only of the mathematical terms but also of the nonlinear phenomena to which they refer. This glossary is not intended to be exhaustive, but rather to highlight a number of important phenomena, so subsequent discussion can be developed with a minimum of confusion as to terminology. In addition to the uniquely nonlinear phenomena, there are a number of concepts from linear processes e.g., equilibrium, stability and resonance, that must be reexamined in this broader context, in order to be made applicable to nonlinear systems. Some of the definitions to follow are taken in part from an unpublished manuscript of the late Dr. Lashivsky.

Active and Passive Systems: An active system contains its own energy source so that it is capable of self-excited oscillation or instability. The rhythmic nature of many biological phenomena, such as the beating of the heart, breathing and some circadian rhythms, to name a few, apparently have this self-excitation property. A passive system contains no internal energy source and thus must be driven by an external force if it is to display any effects of interest. Harvey (1628) was the first to observe that cardiac muscle possessed this self-excitation capability and is therefore an active system.

Dissipation and Restoring Force: We group these two terms together because in a nonlinear system the "dissipation can be either positive or negative. In point of fact the usual situation is one in which the dissipation can change sign during the course of the system's evolution. In a similar way we are interested in system elements which can store energy since these latter terms can be associated with equivalent "restoring forces." We emphasize that there are fundamental differences in the behavior of nonlinear systems with nonlinear dissipation from those with nonlinear restoring forces. An example of a system with a nonlinear dissipation is a van der Pol oscillator. We will not write down the equation here, [cf. Eq. (3.4.1)], but note that the dissipation in such an oscillator is dependent on the amplitude of the oscillator excursion. If the amplitude is low the dissipation is negative and supplies additional energy to the system. If the amplitude exceeds some specific value the dissipation is positive and extracts energy from the system until the amplitude is sufficiently reduced. This oscillator is self-excitatory and was proposed as the basic element in the model of the heart by van der Pol and van der Mark in 1928. This will be discussed more fully when the "relaxation oscillator" is described later in this section.

Entrainment refers to the behavior of a self-excited oscillator with a nonlinear dissipative element subject to a perturbing force at a frequency ω which is approximately the same as the free-running frequency of the oscillator ω_0. Within a narrow range near the free-running oscillation frequency $(\Delta\omega = \omega - \omega_0)$ where $|\Delta\omega| \ll \omega_0$, the frequency of the oscillator is "entrained" to the frequency of the impressed perturbing force, even if the latter is extremely weak. The terms "locked" or "synchronized" are also used to denote the fact that if $|\Delta\omega|$ is less than some value $|\omega_{max} - \omega_0|$ then the $\Delta\omega$ of the oscillator vanishes, i.e., in the entrainment range, $-|\omega_{max} - \omega_0| \leq \omega - \omega_0 \leq |\omega_{max} - \omega_0|$, the response of the oscillator is at frequency ω. Outside this range the oscillator operates at its own frequency and the total response of the system is represented by a combination of the two frequencies. We note that this behavior is strictly nonlinear, being in sharp contrast with the behavior of a linear system. In Figure (3.4.1) the difference between the linear and nonlinear behaviors is depicted. In Figure (3.4.1a) the difference or "beat" frequency between the injected perturbation and the frequency of a linear oscillator is as shown. In Figure (3.4.1b), the beat frequency for a nonlinear oscillator is shown to be zero over a frequency range, the entrainment range, whereas in the linear case only the point of exact resonance has a vanishing beat frequency.

It is also possible for the nonlinear oscillator to be entrained at sub-multiples of the applied frequency, say ω/m, $m = 1, 2, \cdots$ This latter effect was first observed experimentally by van der Pol and van der Mark (1927) and called by them frequency demultiplication. The modern term is subharmonic bifurcation. The particular subharmonic of the applied frequency which entrains the oscillator changes as some parameter of the system is changed. In the 1927 paper this was a capacitance and the authors remark:

"Often an irregular noise is heard in the telephone receivers before the frequency jumps to the next lower value. However, this is a subsidiary phenomenon, the main effect being the regular frequency demultiplication."

The authors were of course mistaken in their judgment regarding that irregular noise, since understanding the cause of that noise has in fact occupied the time of an exponentially growing segment of the scientific community over the past decade or so. The "irregular noise" is now called "chaos" and has been used to describe everything from the fluctuations in the velocity of turbulent fluids to the semi-classical interpretation of probability densities in quantum phenomena to the observed variations in the populations in ecological systems. Chaos appears to be a generic property of nonlinear phenomena and in the next section we will discuss its nature is some detail.

Equilibrium and Stability: In linear systems the term equilibrium is usually applied in connection with conservative forces, with the point of equilibrium

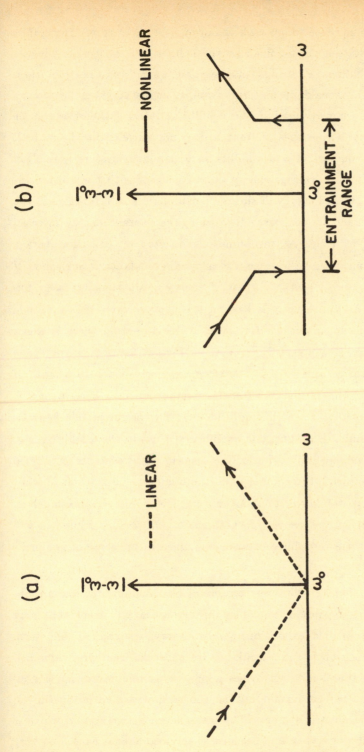

Figure 3.4.1. The difference between the concept of "resonance" in the linear oscillator (a) and the nonlinear (relaxation) oscillator (b) is shown. In (a) only a single point resonance occurs whereas in (b), there is a finite range of frequencies where resonance occurs.

corresponding to the vanishing of all forces with the system being at rest. The stability of such an equilibrium state is then defined by the behavior of the system when it is subject to a small perturbation, i.e., a small displacement away from the equilibrium state in phase space. Roughly speaking, the terms stability or instability indicate that after the perturbation the system returns to the equilibrium state (stable) or that it continues to move away from it (unstable) or that it does not move (neutrally stable). To analyze the notions of equilibrium and stability in a nonlinear system it is often convenient to examine the flow of average energy in an oscillator with nonlinear dissipation Figure (3.4.2). The average energy dissipated per unit time as a function of the oscillation amplitude is denoted by the curve $E^{(-)}$ while the average energy injected into the system is denoted by $E^{(+)}$. At the amplitude denoted by A_0 the average rates of energy input and output are the same and a dynamic balance is achieved. If the amplitude increases, $A_0 \rightarrow A_0 + \delta A_0$, the rate of energy dissipation increases [follow $E^{(-)}$ curve in Figure (3.4.2)]. Energy is lost on the average, and the amplitude returns to A_0. Similarly, if the amplitude decreases, $A_0 \rightarrow A_0 - \delta A_0$, there is a net input of a energy on the average [follow $E^{(+)}$ curve in Figure (3.4.2)] the amplitude grows, and the system again returns to A_0. Therefore, A_0 corresponds to a state of stable equilibrium. However this is a state of stable equilibrium in a *dynamic* sense, since the system executes oscillations with amplitude A_0. The origin in this figure is also a point of equilibrium but it is an unstable equilibrium point in the "static" sense, i.e., in terms of the more familiar application of the term in linear systems mentioned earlier. Note that in Figure (3.4.2) the point A_0 and the origin are both states of what is sometimes called marginal stability in the conventional linear analysis, inasmuch as there is zero energy transfer, on the average, in both cases. On the other hand, it will be evident that these points are very different when examined in the sense of nonlinear systems.

Harmonics and Subharmonics: While the process of harmonic production is probably familiar to the reader, the production of subharmonics is a concept that may require careful definition. In this exposition, unless otherwise qualified, the term subharmonic means the appearance of an oscillatory response at fractional submultiples of an external forcing function which drives a passive system possessing a nonlinear restoring force. This process should not be confused with the production of subharmonic output in a parametric system. The latter is an active system which is capable of self-excited oscillations when an energy-storage parameter is caused to vary periodically by an external agency.

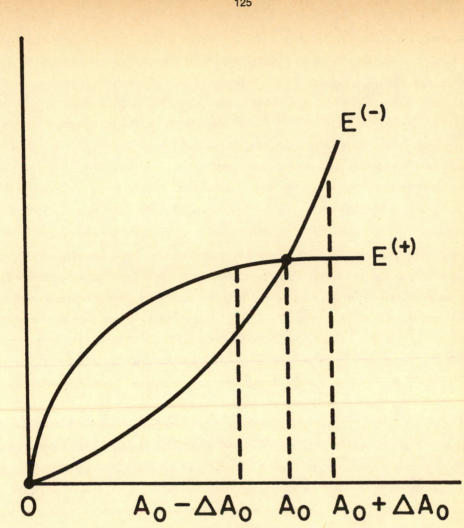

Figure 3.4.2.　　The stability of the point A_o is indicated as the system is displaced both above and below this point. The curve $E^{(+)}$ is the average energy injected into the system and $E^{(-)}$ the average energy extracted from the system, the intersection point is one of dynamic balance. See text for discussion.

Hysteresis: We shall find that a critical difference between linear and nonlinear systems is the fact that in certain cases nonlinear systems are capable of multiple responses to a given forcing function. That is to say, the curve representing the system response versus its forcing function is multiple-valued, implying that the curve is a figure which encloses a finite volume, rather than a curve which retraces itself, as in the case for a single-valued response function. An idealized hysteresis curve is shown in Figure (3.4.3) together with an idealized single-valued curve. Such curves are quite familiar in systems theory where one axis denotes the input to a system and the other axis the output. The finite area enclosed results from the "dissipative" character of the system.

Jump Phenomenon: This term refers to the resonance response of a passive system possessing a nonlinear restoring force. The fact that the restoring force is nonlinear means that it is a function of the amplitude, so that the resonance frequency becomes a function of the strength of the forcing function. A family of response curves corresponding to this situation is shown in Figure (3.4.4). The term "jump phenomenon" describes the effect which arises when the response curve has been distorted to the extent that a vertical tangent becomes possible. In this case the system response function can exhibit hysteresis. Hysteresis here means that if the amplitude of the forcing function is kept constant in tracing a resonance curve (by varying the frequency) then different results are obtained depending on whether the traverse is performed in the direction of increasing frequency or decreasing frequency. In Figure (3.4.4) the case of increasing frequency is indicated by the path a-b-c-d-e while the case of decreasing frequency is indicated by e-d-f-b-a. We note that the jump phenomenon is strictly a manifestation of the nonlinear character of the system.

Limit Cycle, Hard and Soft Excitation: The phase plane of a Hamiltonian (no dissipation) oscillator is shown in Figure (3.4.5a) together with the limit cycle for an oscillator with nonlinear dissipation Figure (3.4.5b). Although there are superficial resemblances between these diagrams, there are, in fact, fundamental differences between these two physical systems. While the linear conservative oscillator can be described by an infinite family of closed ellipses, as shown in (a) of the figure, the nonlinear oscillator approaches a *single limit cycle* as seen in (b). This limit cycle is reached asymptotically whether the initial conditions correspond to an infinitesimal perturbation near the origin or to a finite perturbation far beyond the limit cycle. In both cases the phase point spirals to the limit cycle, which is a stable final state. On the other hand, the conservative linear oscillator does not display this "structural stability." Any perturbation causes it to leave one ellipse and move to another.

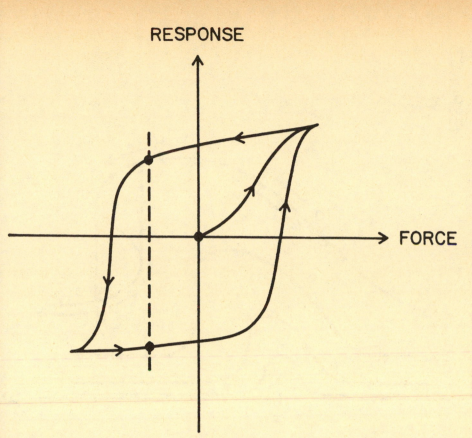

Figure 3.4.3.   Two idealized response-force curves are depicted; the outer curve depicts a system with hysteresis, the inner one a system in which the
response is directly proportional to the force. The hysteresis curve is
multivalued and implies that the system is dissipative, whereas the single valued response curve implies that the system is reversible.

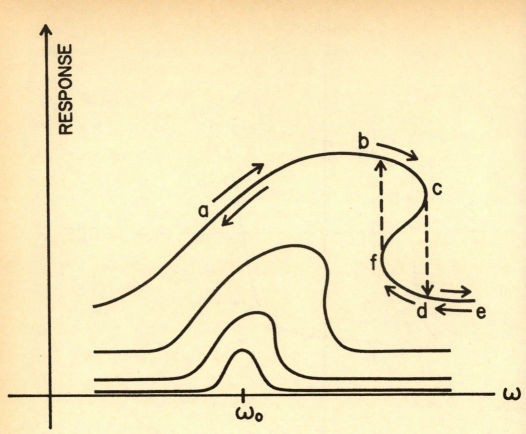

Figure 3.4.4.   A family of resonance response curves as a function of frequency for a passive system possessing a nonlinear restoring force is depicted. The path b-c-d-f-b denotes the "jump phenomenon" described in the text.

The terms "hard excitation" and "soft excitation" are frequently misunderstood and can be explained by reference to Figure (3.4.5). Many nonlinear systems are capable of oscillation in more than one limit cycle, as shown in Figure (3.4.5c). The system approaches a given limit cycle or departs from it, depending on whether the limit cycle is stable or unstable. In the configuration shown in Figure (3.4.5c) the inner limit cycle is unstable and the outer one is stable. Therefore, in order for the system to start from the origin and reach the stable limit cycle the initial position and velocity must correspond to a finite impulse sufficient to carry the system past the unstable limit cycle. This process is called "hard" excitation. On the other hand, the system shown in Figure (3.4.5b) corresponds to a "soft" excitation inasmuch as an infinitesimal perturbation close to the origin will start the system on a trajectory which ultimately reaches the limit cycle.

Mode Competition is a phenomenon characteristic of a system in which a mechanism exists to provide nonlinear coupling between linear modes, recall the FPU problem of heat transport. In contrast with a linear system, in which each mode oscillates independently, in a nonlinear system an exchange of energy is possible between the energy reservoir and the individual modes, and among the modes themselves. Thus, the final steady state amplitude of a given mode cannot be determined independently, as is the case in the linear system, but must be analyzed by considering its coupling to all the other modes. For example, if the system exhibits nonlinear dissipation the modes compete for energy from the energy source so that if one mode grows larger this energy is taken from the other modes which, in turn, must oscillate at a lower amplitude. This was also observed in the discussion of the FPU problem.

Oscillators: A nonlinear oscillator which is "weakly" nonlinear is capable of oscillating at essentially a single frequency and produces a signal which is very low in harmonic content. Here we will call this a weakly nonlinear oscillator, or *sinusoidal* oscillator, an example being the van der Pol oscillator mentioned earlier. Although the output from such an oscillator system is sinusoidal at a single frequency, there are fundamental and crucial differences between such an oscillator and the classical harmonic oscillator, a conservative system which is loss-free. The basic difference is that the nonlinear oscillator can oscillate at one and only one frequency and one and only one amplitude, the amplitude and frequency being interdependent for a given configuration of parameter values. In contrast, the amplitude and frequency are independent in the classical oscillator, which can oscillate at any arbitrary level for a given set of parameter values. These differences are illustrated in the description of

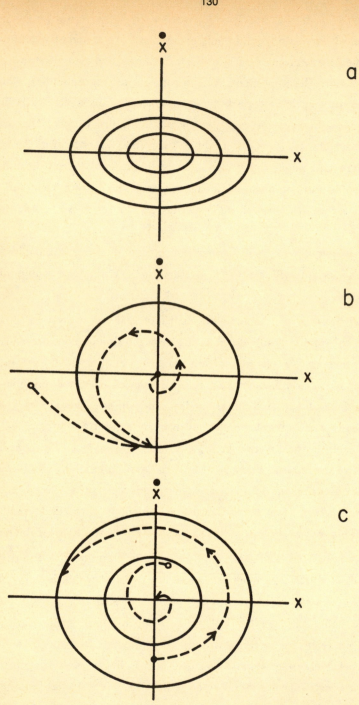

Figure 3.4.5.   A sequence of difference phase space orbits are depicted: (a) neutrally stable limit cycles, distinguished one from the other by the system energy; (b) a stable limit cycle being approached from initial points inside and outside the cycle and (c) "hard" and "soft" excitations (see text).

the limit cycles given above. A typical wave form for a weakly nonlinear oscillator is shown in Figure (3.4.6a).

In contrast with the output of the sinusoidal oscillator, the output wave shape from a *relaxation* oscillator is rich in harmonic content, the wave form exhibiting discontinuities at periodic intervals. Sawtooth or rectangular waveforms are typical. This difference in wave shape reflects some of the basic differences in the two oscillator systems; in certain special cases a given configuration is capable of operation either as a sinusoidal oscillator or as a relaxation oscillator, depending on the values of the parameters, especially those that correspond to the input or loss of energy in the system. It is important to note that the mechanisms which determine the amplitude and frequency of a relaxation oscillator are very different from those in the sinusoidal oscillator. A typical waveform for a relaxation oscillator is shown in Figure (3.4.6b).

3.4.2 The van der Pol Oscillator

For the purpose here we adopt the nomenclature that a biological oscillator is one that is self-excitatory, that is to say, regardless of the initial state of the system the bio-oscillator will approach a stable limit cycle providing that no pathologies arise. As mentioned previously, this idea of an active system was originally proposed in 1928 by van der Pol and van der Mark. They suggested using a nonlinear dynamic equation of the form

$$\ddot{X}(t) + \epsilon[1 - X^2(t)] \ \dot{X}(t) + \omega_0^2 \, X(t) \ = \ 0 \tag{3.4.1}$$

to represent the self-excitation of certain naturally occurring oscillatory systems. In a linear oscillator of frequency ω_0 the coefficient of the first order time derivative determines the stability property of the system. If this coefficient, say λ, is positive then the system is asymptotically stable; i.e., there is a damping $e^{-\lambda t}$ so that the oscillator approaches a fixed point in phase space. If the coefficient λ is negative the solution diverges to infinity as time increases without limit $(e^{\lambda t})$. Of course this latter behavior must terminate eventually since time divergences do not exist in physical systems, at worst stability is lost. But more usually other mechanisms come into play to saturate the growth [cf. §3.2]. In the nonlinear system the coefficient of the "dissipative" term changes sign depending on whether $X^2(t)$ is greater than or less than unity. This property of (3.4.1) leads to a limit cycle behavior of the trajectory in the $(\dot{x},x)$-phase space for the system. The above authors envisioned the application of this limit cycle paradigm to "explain" such phenomena as:

"the aeolian harp, a pneumatic hammer, the scratching noise of a knife on a plate, the waving of a flag in the wind, the humming noise sometimes made

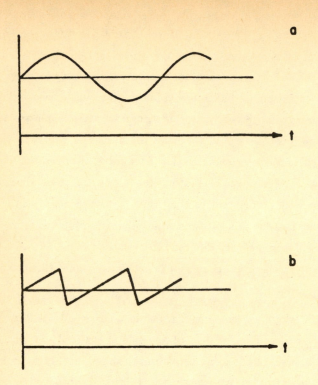

Figure 3.4.6.   Curve (a) is a schematic of a sinusoidal variation in a system. Curve (b) is a schematic of a typical relaxation oscillator time trace.

by a water tap, the squeaking of a door, a neon tube, the periodic re-occurrence of epidemics and of economic crisis, the periodic density of an even number of species of animals living together and the one species serving as food for the other, the sleeping of flowers, the periodic recurrence of showers behind a depression, the shivering from cold, menstruation and, finally, the beating of the heart."

In each of the quoted examples the frequency of the periodic phenomena is determined by some form of a relaxation time. The fundamental period of (3.4.1) was found by van der Pol (1926) to be

$$T_{rel} = 1.61 \frac{\epsilon}{\omega_0^2} \qquad (3.4.2)$$

and since for electrical oscillations $\epsilon/\omega_0^2 = RC$, where R is the resistance and C the capacitance of a simple circuit, this is the relaxation time of the circuit, so the name *relaxation oscillator* was suggested. Thus, (3.4.1) is an example of the generic concept of a relaxation oscillator whose limit cycle behavior was intended to encompass all of the above examples.

There are two limiting modes of oscillation of the solutions to (3.4.1). The first is in the parameter regime $\omega_0^2 \gg \epsilon^2$ where $X(t)$ can start from zero amplitude initially and the self-generating aspect of the system will induce a gradually growing amplitude which finally becomes a constant. This type of growth to a sinusoidal oscillation is depicted in Figure (3.4.7a). The second limiting case is in the parameter regime $\omega_0^2 \ll \epsilon^2$. If the resistance were independent of amplitude then if it is initially positive, this regime would be aperiodic leading to an $X(t)$ that approaches zero asymptotically. However as soon as the instantaneous value of $X^2(t)$ becomes smaller than unity the resistance changes sign and leads to an aperiodic divergence from zero. In Figure (3.4.7b) this condition is seen to lead to a tendency to initially jump away from $X=0$ to a positive value, then to decrease gradually for a longer time and suddenly to jump to a negative value and then to increase slowly for the same length of time. The time period for this rather "square-wave" looking periodicity is given by (3.4.2). Van der Pol and van der Mark argue that this type of periodicity is often found in nature, where the period is not rigidly constant and is given by some form of relaxation time.

Consider for example the aeolian harp, consisting as it does of a string interrupting a flow of air. Behind the string we observe the development of eddies being generated on alternating sides of the wire and diffusing away, thus making room for the

134

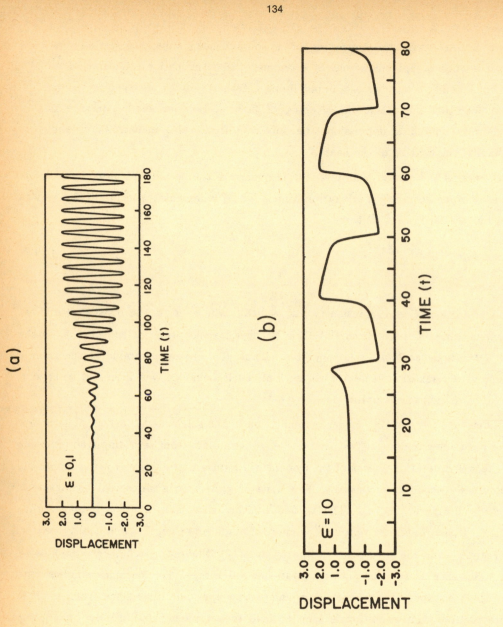

Figure 3.4.7.   The evolution of the van der Pol oscillator for two values of the non-linear strength parameter ϵ is shown.

next eddy to develop, which in its turn diffuses away. Hence the frequency of the sound heard from the Aeolian harp is determined by a diffusion time or relaxation time and is independent of the natural frequency of the string when oscillating in a sinusoidal way. This nonlinear mechanism for sound generation is quite unlike that studied by Euler and d'Alembert in the 19th century that we discussed in §3.1.

One of the properties of the relaxation oscillator is that of *entrainment* and was first discovered by van der Pol and van der Mark (1927). They impressed a small amplitude periodic electromotive force (emf) to an electrical relaxation oscillator. The frequency of the applied emf was initially near resonance, i.e., near the free-running frequency of the relaxation oscillator $2\pi/T_{rel}$. As they varied the relaxation period (T_{rel}) of the oscillator they found that the system continued to vibrate with the period of the impressed emf [cf. Figure (3.4.1)]. The period was varied over an octave of frequencies without change in the system response to the impressed emf. In addition they observed that it was easy to entrain the system to a subharmonic of the impressed frequency ω, i.e., the fundamental frequency of the "entrained" oscillation in the system was ω/n, n being an integer up to 100 or 200. They referred to this phenomenon as *frequency demultiplication,* a term which is no longer in use. The preferred term today is subharmonic bifurcation. Finally the experiment established that the amplitude of the relaxation oscillation could not be considerably influenced by the external emf.

Van der Pol and van der Mark (1928,1929) summarized the properties of the relaxation oscillator as follows:

1. their *time period* is determined by a time constant or relaxation time;

2. their *wave form* deviates considerably from a sine or cosine wave, and, as very steep parts occur, many higher harmonics of pronounced amplitude are present;

3. a small impressed periodic force can easily force the relaxation system to be in step with it (*entrainment* even on subharmonics) while under these circumstances;

4. the amplitude is insensitive to the impressed periodic force;

5. the phenomenon of resonance, so typical for harmonic oscillations, is wholly absent in relaxation oscillations;

6. they were found from a nonlinear differential equation implying the presence of a threshold value resulting in the applicability of the all-or-nothing-law, i.e., either the system is completely active or it is quiescent.

---

[1]In general b is complex but we do not consider that case here [see e.g., Gnedenko and Kolmogorov (1955)].

4. HOW TO BE NONLINEAR

In the previous sections we have repeatedly indicated how linear concepts collapse under the weight of uninterpretable data and are replaced by nonlinear ones. Examples include the existence of non-integer exponents in the functional representation of data, long tails in the distribution functions, saturation phenomena and rhythmic behavior in open (biological) systems. Each has obliged us to seek a nonlinear representation of the underlying process. In each case we observed how a comfortable linear notion was inadequate to describe a particular phenomenon due to the existence of one or more nonlinear mechanisms and therefore had to be discarded. An example we considered is the amplification mechanism responsible for the log-normal distribution being replaced by a Pareto tail at the high end of the income distribution [cf. (§3.2)]. The log-normal distribution relies on a linear interpretation in terms of the logarithm of an individual's income. The amplification mechanism at high incomes is nonlinear, however, and forces the replacement of the linear-based log-normal distribution with the nonlinear-based Pareto power law. The practical question remains: "How does one learn the techniques that are of value in one's field without becoming burdened with perhaps interesting but nonetheless irrelevant mathematical baggage?"

In this section we undertake the task of presenting the salient features of what has transpired in the development of nonlinear dynamics and the understanding of nonlinear processes in general over the past few decades. It would be a gross understatement to say that the discussion is not exhaustive as to the breadth of topics considered, even leaving aside the question of depth in any particular topic. In fact writing this section is much like strolling into a forest and commenting on the various trees that come into view. Some are dramatic, but not of much practical value; others are made of hard wood and useful for building, even others are soft and good for whittling. However, the forest soon becomes dark and impenetrable so we will stand near the forest's edge, from where not too much detail is visible, and scan the wilderness. Hopefully you will find the view as spectacular as I do.

The operational answer to the question of learning only those mathematical techniques of relevance to one's own field of studies to form centers of excellence in which collaboration across traditional disciplinary lines is taken as the norm. It may be argued that such centers already exist. However I would disagree and emphasize that such centers have in the past placed a premium on mathematical techniques and not on the *communication* of those techniques within an applied context. The nonmathematical practitioners must learn the strategy behind the mathematics in order to determine which style of thought will be the most advantageous for their own fields.

In the interim, between what exists today and the establishment of the scientific utopia where disciplinary lines are viewed as quaint anachronisms, we present a menu of phenomena that may suggest either application to, or further development within, a natural philosophy context.

To set the stage for the phenomena to be discussed in this section we note that it has often been commented on that one of the fundamental laws of physics, the second law of thermodynamics, formulated by the 19th century physicist Boltzmann, stands in marked contrast to Darwin's theory of evolution. In fact, all biological and sociological systems evolve from states of lower order to states of higher complexity, in apparent contradiction to the thermodynamic notion that the universe tends toward disorder. The resolution of this paradox has been made a number of times with the observation that the entropy maximization principle upon which the tendency towards disorder is based is a global statement and does not apply to local subsystems such as cells. As Lotka (1925) points out in his remarkable book:

"..... the steady states with which we are most frequently and most closely concerned in the field of organic evolution (our main topic here) are of the second class; not true equilibria in the dynamic sense in which all forces are balanced, but what we have termed quasi-equilibria, states maintained constant or approximately so with a continual expenditure, a continual dissipation or degradation of available energy."

Also from the lectures delivered at the Institute in Dublin in 1943 we quote E. Schrödinger:

"How would we express in terms of the statistical theory the marvelous faculty of a living organism by which it delays the decay into thermodynamical equilibrium (death)? We said before: 'It feeds upon negative entropy,' attracting as it were, a stream of negative entropy upon itself, to compensate the entropy increase it produces by living and thus to maintain itself on a stationary and fairly low entropy level."

Schrödinger points out that his physics colleagues were not very receptive to the idea of a *negative entropy*, in fact the concept was met with "doubt and opposition" in the early 1940's.

The most influential developments for understanding local organization along these lines were made by I. Prigogine and his co-workers over a period of three decades beginning from the middle 40's and for which Prigogine received the 1977 Nobel Prize in Chemistry. He was awarded the prize for "his contributions to nonequilibrium thermodynamics, particularly the theory of dissipative structures." The area of investigation involved nonlinear chemical systems in which the nonlinearity is obtained by feedback loops in reaction mechanisms driven far from equilibrium. Prigogine and his co-workers found that rather than the expected relaxation of such systems towards an equilibrium state these systems would undergo radically different behavior and they obtained a number of unexpected results. For example, as the system is driven far from equilibrium, it may become unstable and then evolve spontaneously to new structures showing coherent behavior. As mentioned by Procaccia and Ross (1977) in their review of the 1977 Nobel Prize in Chemistry;

> "Prigogine refers to the equilibrium and near-equilibrium states as the thermodynamic branch, whereas the new structures are called dissipative structures ..... The dissipative structures can be maintained only through a sufficient flow of energy and matter. The work required to maintain the system far from equilibrium is the source of the formation of order. Fluctuations play a crucial role near the point of instability ..... they become large on the macroscopic scale and are built up by the nonlinear behavior of the system into dissipative structures."

The commentary of Procaccia and Ross goes on to emphasize the general applicability of these concepts outside the domain of irreversible thermodynamics. The concept of a dissipative structure provides a model construct for the analysis of such difficult problems as population dynamics, meteorology, chemical engineering, economics etc. and has been actively developed by others, see e.g. Horsthemke and Lefever (1984).

This general theory of dissipative structures is not of direct concern to us here except insofar as its existence provides us with an analog context for the development of useful ideas in other areas of natural philosophy. For this reason we present a sequence of brief discussions of nonlinear systems, some for their historical importance and others because of their clear experimental vindication of the analysis, but always with the intent that a useful concept will be made available to the reader. Many of the mathematical details are not presented in these discussions in order that more space may be devoted to the interpretative aspects of the problems. However, a modest bibliography is provided for the interested reader.

4.1 Deterministic Chaos

Juxtaposing the words deterministic and chaos, the former indicating the property of determinability (predictability) and the latter that of randomness (unpredictability), usually draws an audience. The expectation of people is that they will be entertained by learning how the paradox is resolved. Here instead we show how the two concepts, long thought to be mutually exclusive, may in fact be the *wave-particle duality*[1] of nonlinear dynamics. From the point of view of dynamics the idea of randomness or fluctuations has traditionally been associated with the (weak) interaction of an observable with the rest of the universe [see §2.3]. The traditional view requires there to be many (an infinite number) degrees of freedom that are not directly observed, but whose presence is manifest through the fluctuations. More recently it has been learned that in a nonlinear system with even a few degrees of freedom chaotic motion can be observed.

What we present in this subsection are some of the recent results obtained in nonlinear dynamics that lead to chaos. First we briefly review the classical work of Lorenz (1963) on a deterministic continuous dissipative system. The phase space orbit for the solution to the Lorenz system is an attractor, but of a kind on which the solution is aperiodic and therefore *strange*. We discuss this family of aperiodic solutions and discover that chaos lurks in a phase space of dimension three. After seeing the brand of chaos that a continuous strange attractor gives we examine a one-dimensional noninvertible nonlinear map. This mapping is the discrete analog of the logistic equation and leads to a subharmonic bifurcation of the solution eventually resulting in a kind of chaos that is distinct from that generated by the Lorenz attractor. The results are quite general for maps with a single maxima. A third kind of chaos is related to that found on the Lorenz attractor in that a two-dimensional invertible map is shown to have an attractor set which is *strange*, i.e., it is the discrete analog of the Lorenz attractor.

One of the fascinating aspect of these maps is that they appear to be the natural way in which to describe the time development of systems in which successive generations are quite distinct. Thus they are appropriate for describing the change in population levels between successive generations: in biology, where populations can refer to the number of individuals in a given species or the gene frequency of a mutation in an evolutionary model; in sociology, where population may refer to the number of people adopting the latest fad or fashion; in medicine, where the population is the number of individuals infected by a contagious disease; and so on. The result of the mathematical analysis is that for certain parameter regimes there are a large number of classes of discrete dynamical models (maps) with chaotic solutions. The *chaos* associated with these solutions is such that the orbits are aperiodic or erratic in time, but the chaos of one class has not been shown to be the same as that of another class. However they all indicate that one must abandon the notion that a deterministic nonlinear map of a process implies a predictable result. One may be able to solve the discrete equations of motion only to find a chaotic solution that requires a distribution function for making predictions.

4.1.1 Continuum Systems (New Chaos)

The new view of randomness that we have been discussing came about through the attempts of E. Lorenz (1963) to understand the short term variability of weather patterns and thereby enhance the predictability of the weather. His approach was to represent a forced dissipative geophysical hydrodynamic flow by a set of deterministic nonlinear differential equations with a finite number of degrees of freedom. For the particular physical problem he was investigating, the number of degrees of freedom he was eventually able to use was three, let's call them X, Y and Z. In the now standard form these equations are

$$\frac{dX}{d\tau} = -\sigma X + \sigma Y \qquad (4.1.1a)$$

$$\frac{dY}{d\tau} = -XY + rX - Y \qquad (4.1.1b)$$

$$\frac{dZ}{d\tau} = XY - bZ \qquad (4.1.1c)$$

whose solutions can be identified with trajectories in phase space.[2] What is of interest here are the properties of nonperiodic bounded solutions in this three dimensional phase space. A bounded solution is one that remains within a restricted domain of phase space given that it was in that domain initially.

The phase space for the set of equations (4.1.1), is three-dimensional and the solution to these equations trace out a curve $\Gamma_t(x,y,z)$ given by the locus of values of $\mathbf{X}(t) = (X(t), Y(t), Z(t))$. We can associate a small volume $V_0(t) = X_0(t)\, Y_0(t)\, Z_0(t)$ with the perturbation of the trajectory and investigate how this volume of phase space changes with time. If the original flow is confined to a region R then the rate of change of the small volume with time $\partial V_0/\partial t$ must be balanced by the flux of volume $\mathbf{J}(t) = V_0(t)\, \mathbf{X}(t)$ across the boundaries of R. The quantity $\mathbf{X}(t)$ in the flux $\mathbf{J}$ represents the time rate of change of the dynamical variable in the absense of the perturbations, i.e., the unperturbed flow field that can sweep the perturbation out of the region R. The balancing condition is expressed by an equation of continuity and in the physics literature is written

$$\frac{\partial}{\partial t}\, V_0(t) + \nabla \cdot \mathbf{J}(t) = 0 \qquad (4.1.2)$$

or in detail

$$\frac{1}{V_0(t)}\, \frac{d}{dt}\, V_0(t) = \partial_x \dot{x} + \partial_y \dot{y} + \partial_z \dot{z} \qquad (4.1.3)$$

where $\frac{d}{dt}\, (\equiv \partial_t + \dot{\mathbf{x}} \cdot \nabla_x)$ is the so-called *convective* or *total* derivative of the volume. Using the equations of motion (4.1.1) in (4.1.3) we obtain

$$\frac{1}{V_0(t)} \; \frac{d}{dt} \; V_0(t) \; = \; -(\sigma + b + 1) \quad . \tag{4.1.4}$$

Equation (4.1.4) is interpreted to mean that as an observer moves along with an element of phase space volume $V_0(t)$ associated with the flow field, the volume will contract at a rate $(b + \sigma + 1)$, i.e., the solution to (4.1.4) is $V_0(t) = V_0(t=0) \exp\{ -(b + \sigma + 1)t\}$. Hence the volume goes to zero as $t \rightarrow \infty$ at a rate which is indepedent of the solutions $X(t)$, $Y(t)$ and $Z(t)$. As pointed out by Lorenz (1963), this does not mean that each small volume shrinks to a point in phase space; it may simply become flattened into a surface. Consequently the total volume of the region initially enclosed by the surface R shrinks to zero at the same rate, resulting in all trajectories becoming asymptotically confined to a specific subspace having zero volume.

To understand the relation of a system with the kind of dynamical situation we were discussing earlier we must study the behavior of the system on the limiting manifold to which all trajectories will be ultimately confined. This cannot be done analytically because of the nonlinear nature of the equations of motion (4.1.1). Therefore, these equations are integrated numerically on a computer and the resulting solution is depicted as a curve in phase space for particular values of the parameters σ, b and r. The technical details associated with the mathematical understanding of these solutions is available in the open literature, see e.g., Ott (1981) or Eckmann (1981).

In Figure (4.1.1) we display the behavior of $Y(t)$ for 3000 time units. After reaching an early peak at $t=35$, $Y(t)$ relaxes to a relatively stable value at $t=85$ which persists, subject to systematically amplified oscillations, until near $t \simeq 1650$. Beyond this time $Y(t)$ becomes pulse-like and appears to change signs at apparently random intervals. This irregularity is not just in the spacing between maxima but also in the sign of the adjacent maxima, i.e., the irregular occurrance of a number of peaks of one sign before a peak of the apposite sign occurs.

In Figure (4.1.2a) the solution manifold in the three dimensional phase space is shown and (4.1.2b) projects of the solution manifold onto the (z,y)-plane and the (x,y)-plane. The trajectory indicated is not complete, but is that segment traversed in the time interval $t = 1400$ to 1900. The points C and C' are the fixed points of the equations, i.e., the values of x,y and z for which $\dot{X} = \dot{Y} = \dot{Z} = 0$ in (4.1.1). These two views of the trajectory indicate that the erratic behavior apparent in the $Y(t)$ plot arises from the orbit spiraling around one of the fixed points C or C' for some arbitrary period and then jumping to the vicinity of the other fixed point, spiraling around that for a while and then jumping back to the other and on and on. Virtually

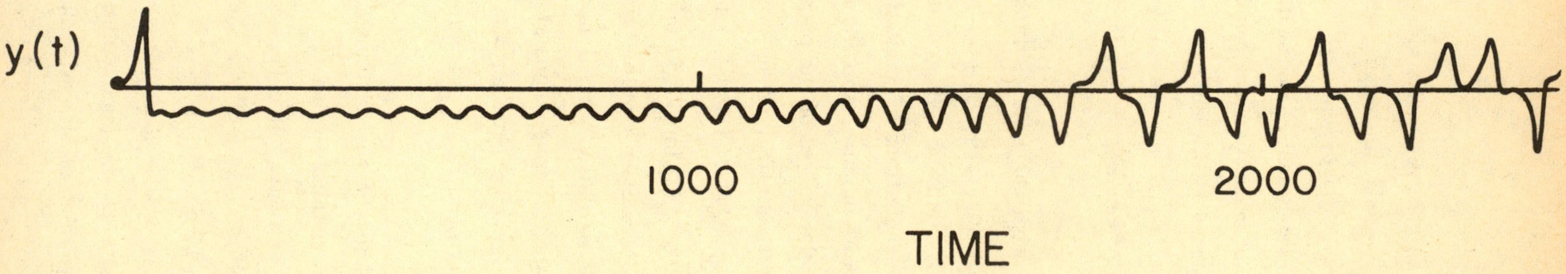

Figure 4.1.1. The time history of the Y(t) component of the solution to the Lorenz system of equations is shown for 3×10^3 time units [from Lorenz (1963)].

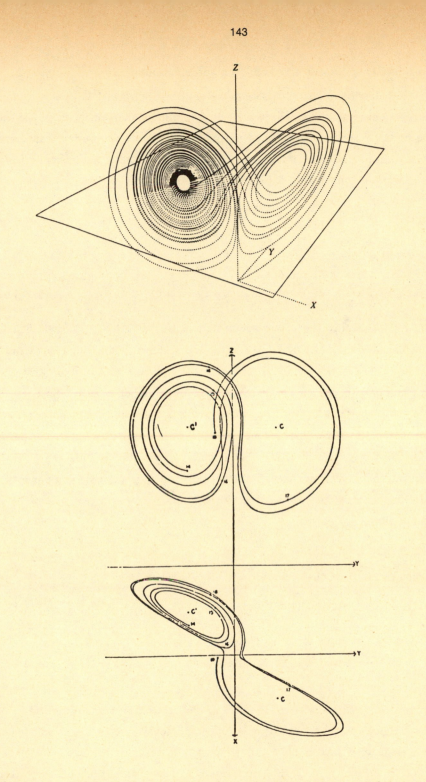

Figure 4.1.2.   The three-dimensional attractor solution for the Lorenz system is depicted in (a). In (b), the projections of the attractor onto z - y and x - y planes, respectively, are shown [from Lorenz (1963)].

all trajectories finally end up on this highly unstable manifold.

The erratic behavior in the time series depicted in Figure (4.1.1) is also apparent in the associated spectrum. The spectrum, as you will recall from §2.2, is the mean square value of the Fourier transform of a time series, i.e., the Fourier transform of the correlation function. Consider the solution $X(t)$; it will have a Fourier transform over a time interval T defined by [cf. (§2.2)]

$$\hat{X}_T(\omega) = \int_0^T X(t) e^{-i\omega t} \frac{dt}{2\pi} \tag{4.1.5}$$

and a power spectral density (PSD)

$$S_{xx}(\omega) \equiv \lim_{T \to \infty} \frac{|\hat{X}_T(\omega)|^2}{T} . \tag{4.1.6}$$

In Figure (4.1.3) we display the power spectral densities (PSD) $S_{xx}(\omega)$ and $S_{zz}(\omega)$ as calculated by Farmer et al. (1980) using the trjectory shown. It is apparent from the $S_{xx} - PSD$ that there is no dominant periodic x-component to the dynamics of the attractor, although lower frequencies are favored over higher ones. The $S_{zz} - PSD$ has a much flatter spectrum overall, but there are a few isolated frequencies at which energy is concentrated. This energy concentration would appear as a strong periodic component in the time trace of $Z(t)$. From this one would conclude that $X(t)$ is non-periodic, but that $Z(t)$ possesses both periodic and non-periodic components. In fact from the linearity of the Fourier transform (4.1.5) we would say that $Z(t)$ is a superposition of these two parts:

$$Z(t) = Z_p(t) + Z_{np}(t) . \tag{4.1.7}$$

The implication of (4.1.7) is that the auto-correlation function

$$C_{zz}(\tau) = \lim_{t \to \infty} <Z(t)Z(t+\tau)> \tag{4.1.8}$$

may be written as the sum of a nonperiodic component $<Z_{np}(t)Z_{np}(t+\tau)>$ that decays to zero as $\tau \to \infty$ and a periodic component $<Z_p(t)Z_p(t+\tau)>$ that does not decay.

To summarize: We have here a new kind of attractor that is referred to as "strange" or "chaotic" in that the power spectral density resulting from the time series of the trajectory has broad-band components. Dynamical systems that are periodic or quasi-periodic have a PSD composed of delta functions, i.e., very narrow spectral peaks; non-periodic systems have broad spectra with no dramatic emphasis of any particular frequency. It is this broad band character of the PSD that is currently used to identify non-periodic behavior in experimental data.

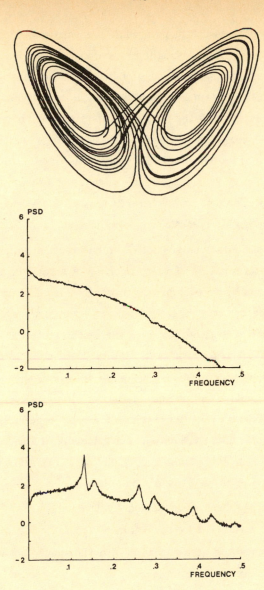

Figure 4.1.3. The projection of the Lorenz attractor onto the x-z plane is shown in (a). The power spectral density $S_{xx}(\omega)$ and $S_{zz}(\omega)$ is calculated using this solution for $x(t)$ and $z(t)$ [Farmer et al. (1982)] and depicted in (b) and (c), respectively.

So what does this all mean? In part what it means is that a sytem such as our earlier example of the stock market might be stochastic[3] even if its dynamics could be "isolated" to a few (three or more) degrees of freedom that interact in a determinstic but nonlinear way. If the system is dissipative, i.e., energy (money) is extracted from the system on the average, but the system is open to the environment, i.e., energy (money) is supplied to the system by means of a boundary condition, then a "strange attractor" is not only a possible manifold for the solutions to the dynamic equations; it, or something like it, may even be probable. Haken (1978; pp. 320) has shown that a set of equations of precisely the same structure as (4.1.1) arises in laser physics and therefore has an analogous interpretation.

Of course these considerations are not of much practical value unless they can be implemented in the determination of the properties of a real data set. The rationale for their application was also developed by Lorenz in his seminal work, but the full extent of its importance has only recently begun to emerge, see eg. Lanford (1976). He (Lorenz) observed that the trajectory leaves the spiral centered at C say, only after exceeding some critical distance from the center. Further, the degree to which this critical distance is exceeded determines the point at which the next spiral, i.e., that centered at C', is entered as well as the number of circuits executed prior to making the transition back to the C center again. Thus he concludes that "some single feature of a given circuit should predict the same feature of the following circuit." As an example he selected the maximum value of the $Z(t)$ variable along the trajectory which occurs whenever the circuit is nearly completed.

In Figure (4.1.4) the abscissa is labeled by the value of the *nth* maxima Z_n of $Z(t)$ and the ordinate is labeled by the value of the following maximum Z_{n+1}. It is clear that the point generated lie along a curve if the spaces between points are filled in. This is shown for example by Shaw (1981) using the increased computing capacity that had developed in the intervening years. The computer generated function clearly prescribe a two-to-one relation between Z_n and Z_{n+1}. From this relation one could formulate an empirical prediction scheme using the geometry of the attractor as a data set without a knowledge of the underlying dynamical equations. This has recently been done by Simoyi, Wolf and Swinney (1982) for a different system but one in which successive maxima of a dynamic variable are related by a one-dimensional map. The experiments they conducted were on the Belousov-Zhabotinskii reaction in a stirred flow reactor and the dynamic variable of interest is a single chemical concentration. The experimentally determined relation is depicted in Figure (4.1.5).

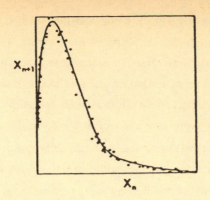

Figure 4.1.4.   Corresponding values of relative maximum of Z (abscissa) and subsequent relative maximum of Z (ordinate) occurring during the first 6000 iterations. [From Lorenz (1963)].

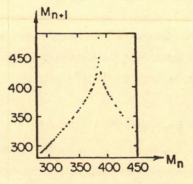

Figure 4.1.5.   A single chemical species in the well-stirred Belousov-Zhabotinskii reaction has been shown to have a trajectory lying on a strange attractor in a three-dimensional phase space. The Poincaré surface of section of this attractor yields the indicated one-dimensional mapping function directly from the data. [Taken from Simoyi et.al. (1982)].

Some of these ideas have recently been clarified and extended by a number of scientists among whom are Packard, Crutchfield, Farmer and Shaw (1980). These latter investigators have used time series to reconstruct the phase space geometry of a dynamical system. The underlying notion of the method of reconstruction is that if the state of a system is completely described at time t by n variables $[X_1(t), X_2(t), ..., X_n(t)]$ then any other set of n "independent" variables should provide an equivalent characterization of the system. The quantities ordinarily used are the values of the phase space coordinates, which for the Lorenz attractor are the triad $[X(t), Y(t), Z(t)]$. A different three-dimensional system, one constructed by Rössler (1976) to describe a certain chemical reaction was considered by Packard et al. (1980). They constructed a time series by sampling a single coordinate of this three-dimensional system and found a variety of three independent quantities which produced a faithful phase-space representation of the original dynamics. One set they considered is $[X(t), \dot{X}(t), \ddot{X}(t)]$ from which they constructed the phase space attractor in the $(x, \dot{x})$-plane and found it to be topologically equivalent to a two-dimensional projection of the original attractor. Another set of three quantities is that obtained from $X(t)$ displaced from itself by a specified internal τ, i.e., the triad $[X(t), X(t-\tau), X(t-2\tau)]$. This latter technique is the one employed by Simoyi et al. (1982) in their study of the Belousov-Zhabotinskii reaction and has the advantage of not introducing additional *noise* into the data set. Taking derivatives of experimental time series which would be required in the $[x, \dot{x}, \ddot{x}]$ representation would generate significant noise because of the rapid changes in $X(t)$ with time.

4.1.2 Discrete Systems (Another New Chaos)

The kind of chaos we discussed in the preceding section resulted from the dynamical properties of continuous ordinary differential equations describing dissipative systems open to the environment. The chaos is a result of the system's sensitivity to initial conditions. In this section we examine the evolution of systems that can be described by discrete times and which are seen by many as an attempt to isolate simplifiying features of dissipative dynamical systems, see e.g. Collet and Eckmann (1980). The evolution equation in such a discrete representation is called a *map* and the evolution is given by iterations of the mapping. Thus iterations of the form $x_n \rightarrow x_{n+1} = f(x_n)$, where f maps the interval $[0, 1]$ into itself, is interpreted as a discrete time version of a continuous dynamic system. The choice of interval $[0,1]$ is arbitrary since the change of variables $y = (x-a)/(b-a)$ will replace a mapping of the interval $[a, b]$ into itself by one that maps $[0,1]$ into itself. For example, consider the continuous trajectory in the two-dimensional space depicted in Figure (4.1.6). The

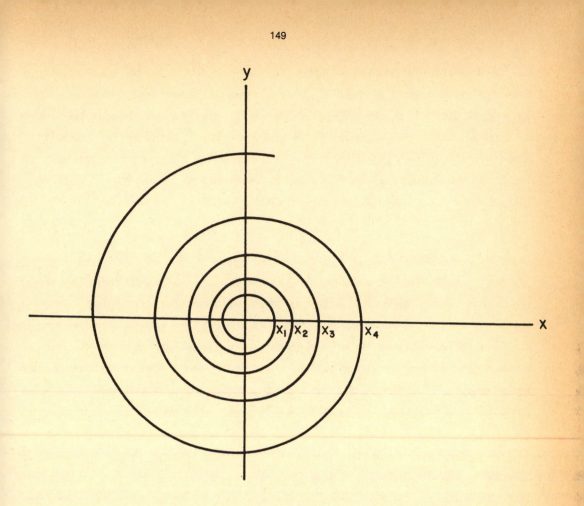

Figure 4.1.6.   The spiral is an arbitrary orbit depicting a function $y = f(x)$. The intersection of the spiral with the x-axis defines a set of points $x_1, x_2, ...,$ that can be obtained from a mapping determined by $f(x)$.

intersection points of the orbit with the x-axis are denoted by x_1, x_2,.... The point x_{n+1} can certainly be related to x_n by means of a function f determined by the trajectory. Thus instead of solving the continuous differential equations that describe the trajectory, in this approach one produces models of the mapping function f and studies the properties of $x_{n+1} = f(x_n)$. Here n plays the role of the time variable. This strategy has been applied to models for biological, chemical and physical systems. May (1976) has pointed out a number of possible applications of the the fundamental equation

$$x_{n+1} = f(x_n) \tag{4.1.9}$$

In genetics, for example, x_n would describe the change in the gene frequency between successive generations; in epidemology, the variable x_n could denote the fraction of the population infected at time n; in psychology, certain learning theories can be cast in this form where x_n is interpreted as the number of bits of information that can be remembered up to generation n; in sociology, x_n might be interpreted as the number of people having heard a rumor at time n and (4.1.9) would then describe the propagation of rumors in societies of various structures, e.g., Kemeny and Snell (1972). The potential applications of this equation are therefore restricted only by our imaginations.

To relate some of the ideas that we will introduce shortly to those that we have already discussed it is useful to have gone over the new ideas once in a specific context. Therefore we will reconsider the process of growth, only now using discrete equations rather than the continuous rate equations of §3.3. These ideas are also well described by Segal (1984) in a biological dynamics context and we borrow his format for the next few paragraphs.

Consider the simplest mapping, also called a recursion relation, in which the population x_n of organisms per unit area in the *nth* generation is strictly proportional to the population in the preceding generation with a proportionality constant μ:

$$x_n = \mu x_{n-1} \quad , \quad n = 1, 2, \dots \tag{4.1.10}$$

The proportionality constant is given by the difference between the birth rate and death rate and is therefore the *net* birth rate of the population. Equation (4.1.10) is quite easy to solve. Suppose that the population has a level $x_0 = N_0$ at the initial generation, then the recursion relation yields the sequence of relation

$$x_1 = \mu N_0 \quad , \quad x_2 = \mu x_1 = \mu^2 N_0, \cdots \tag{4.1.11}$$

so that in general

$$x_n = \mu^n N_0, \quad n = 0, 1, \cdots \tag{4.1.12}$$

This rather simple solution already exhibits a number of interesting properties. Firstly, if the net birthrate μ is less than unity, then we can write $\mu^n = e^{-n\beta}$ where $\beta > 0$, so that the population decreases exponentially between successive generations (note $\beta = -\ln\mu$). This is a reflection of the fact that with $\mu < 1$, the insect population fails to reproduce itself from generation to generation and therefore it exponentially approaches extinction:

$$\lim_{n \to \infty} x_n = 0 \quad \text{if} \quad \mu < 1 \quad . \tag{4.1.13a}$$

On the other hand if $\mu > 1$, then we can write $\mu^n = e^{n\beta}$ where $\beta (= \ln\mu) > 0$, so the population increases exponentially between successive generations. This is a reflection of the fact that with $\mu > 1$ the population has an excess at each generation resulting in a population explosion. This is the Malthus' exponential population growth:

$$\lim_{n \to \infty} x_n = \infty \quad \text{if} \quad \mu > 1 \quad . \tag{4.1.13b}$$

The only value of μ for which the population does not have these extreme tendencies is $\mu = 1$, when since the population reproduces itself exactly in each generation we obtain the unstable situation:

$$\lim_{n \to \infty} x_n = N_0 \quad \text{if} \quad \mu = 1 \quad . \tag{4.1.13c}$$

Of course this simple model is no more valid than the continuous growth law of Malthus (3.3.1). The arguments in §3.3 that allowed us to generalize that description to include the effects of limited resources in the growth law apply equally well in this discrete context. In particular the arguments of Verhulst still apply to saturate the growth of the population in the later generations. One way of viewing the change in the functional form of the growth law proposed by Verhulst (3.3.3) is that the birth rate μ becomes population dependent. In particular, the birthrate is assumed to decrease with increasing population in a linear way:

$$\mu \to \mu(x_n) = \mu[1 - x_n/\Theta] \tag{4.1.14}$$

where Θ is the saturation level of the population. Thus the linear recursion relation (4.1.10) is replaced with the nonlinear discrete *logistic equation,*

$$x_{n+1} = \mu \, x_n \, [1 - x_n/\Theta] \quad . \tag{4.1.15}$$

It is clear that when $x_n \ll \Theta$ the population grows exponentially since the nonlinear term is negligible. However at some point the ratio x_n/Θ is going to be of the order unity and the rate of population growth will be retarded. When $x_n = \Theta$ there are no more births in the population. Biologically the regime $x_n > \Theta$ corresponds to a negative birthrate, but this does not make biological sense and so we restrict the region of *interpretation* of this model to $[1 - x_n/\Theta] > 0$. Finally, we reduce the number of parameters from two, μ and Θ, to one by introducing $y_n = x_n/\Theta$ the fraction of the saturation level achieved by the population. In terms of this *ratio* variable the recursion relation (4.1.15) is

$$y_{n+1} = \mu \, y_n [1 - y_n] \quad . \tag{4.1.16}$$

Segal (1984) challenges the readers of his book (at this point) to attempt and predict the type of behavior manifest by the solution to (4.1.16), eg. Are these periodic components to the solution? Does extinction ever occur?, etc. His intent was to alert the reader to the inherent complexity contained in the deceptively simple looking equation (4.1.16). We will examine some of these general properties shortly, but first let us explore our example a bit more fully.

We noticed that extinction was the solution to the simple system (4.1.10) when $\mu < 1$. Is extinction a possible solution to (4.1.16)? If it is, then once that state is attained, it must remain unchanged throughout the remaining generations. Put differently, extinction must be a steady-state solution of the recursion relation. Let us assume the existence of a steady-state level y_{ss} of the population such that (4.1.16) becomes

$$y_{ss} = \mu \, y_{ss} (1 - y_{ss}) \tag{4.1.17}$$

for all n. Equation (4.1.17) defines the algebraic equation

$$y_{ss}^2 + \left(\frac{1}{\mu} - 1 \right) + y_{ss} = 0 \tag{4.1.18}$$

which has the two roots $y_{ss} = 0, (1 - \frac{1}{\mu})$. The $y_{ss} = 0$ root is extinction, but we now have a second steady solution to the mapping, $y_{ss} = 1 - \frac{1}{\mu}$. One of the questions we will address in the more general treatment of this problem is to determine to which of these steady states the population evolves as the years go by.

Before we examine the more general properties of (4.1.16) and equations like it, let us use a more traditional tool of analysis and examine the stability of the two steady states found above. Traditionally the stability of a system in the vicinity of a given value is determined by perturbation theory. We use that technique now and write

$$y_n = y_{ss} + \xi_n \tag{4.1.19}$$

where $\xi_n \ll 1$ so that (4.1.19) denotes a small change in population from its steady state value. If we now substitute (4.1.19) into (4.1.16) we obtain

$$y_{ss} + \xi_{n+1} = \mu(y_{ss} + \xi_n)[1 - y_{ss} - \xi_n] \quad . \tag{4.1.20a}$$

Then using (4.1.17) to eliminate certain terms and neglecting terms quadratic in ξ_n we obtain

$$\xi_{n+1} \simeq (\mu - 2y_{ss})\xi_n \tag{4.1.20b}$$

as the recursion relation for the perturbation. In the neighborhood of extinction, the $y_{ss} = 0$ steady state, (4.1.20b) reduces to (4.1.10). Therefore if $0 < \mu < 1$ then the fixed point $y_{ss} = 0$ is stable and if $\mu > 1$ the fixed point is unstable. By stable we mean that $\xi_n \to 0$ as $n \to \infty$ if $0 < \mu < 1$ so that the system returns to the fixed point and by unstable we mean that $\xi_n \to \infty$ as $n \to \infty$ if $\mu > 1$ so that the perturbation grows without bound and never returns to the fixed point. Of course $\mu = 1$ means the fixed point is neutrally stable, i.e., it neither return to nor diverges from $y_{ss} = 0$.

In the neighborhood of the steady state $y_{ss} = 1 - \dfrac{1}{\mu}$ the recursion relation becomes

$$\xi_{n+1} = (2 - \mu)\xi_n \quad . \tag{4.1.21}$$

The preceding analysis can again be repeated with the result that if $1 > 2 - \mu > -1$ the fixed point $y_{ss} = 1 - \dfrac{1}{\mu}$ is stable and implies that the birthrate is in the interval $1 < \mu < 3$. The stability is monotonic for $1 < \mu < 2$, but because of the changes in sign it is oscillatory for $2 < \mu < 3$. Similarly the fixed point is unstable for $0 < \mu < 1$ (monotonic) and $\mu > 3$ (oscillatory).

Returning now to the more general context it may appear that limiting the present analysis to one-dimensional systems is unduly restrictive; however, we recall that the system is pictured to be a projection of a more complicated dynamical system onto a one-dimensional subspace [cf. e.g., Figure (4.1.6)]. This is not unlike the argument we made earlier for the observable leading to the Langevin equation [cf. §2.3]; however the observable here is a discrete rather than a continuous quantity and the resulting "dynamic" equation (4.1.9) is deterministic rather than stochastic. A substantial literature based on (4.1.9) has developed in the past decade, much of which is focused on the purely mathematical properties of such mappings. We are not concerned with that vast literature here, except insofar as it makes available to us solutions and insights that can be applied in natural philosophy. The physicists have been quite actively exploring the consequences of these results for physical and chemical systems, but with a few notable exceptions activity in the other sciences has been relatively subdued.

One of the exceptions alluded to is the remarkable review article of May (1976) in which he makes clear the state of the art in discrete systems up until that time. In addition he makes the following comments:

> "The review ends with an evangelical plea for the introduction of these difference equations into elementary mathematics courses, so that students intuitions may be enriched by seeing the wild things that simple nonlinear equations can do."

His plea was motivated by the recognition that the traditional mathematical tools such as Fourier analysis, orthogonal functions, etc. are all fundamentally linear and

> "...the mathematical intuition so developed ill equips the students to confront the bizarre behavior exhibited by the simplest discrete nonlinear systems, ... Yet such nonlinear systems are surely the rule, not the exceptions, outside the physical sciences."

May ends his article with the following indictment:

> "Not only in research, but also in the everyday world of politics and economics, we would all be better off if more people realized that simple systems do not necessarily possess simple dynamic properties."

We shall make the assumption for the remainder of this section that the maps (dynamic systems) of interest contain a single maximum and that $f(x)$ is monotonically increasing for value of x below this maximum and monotonically decreasing for values of x above this maximum. Maps such as these, i.e., maps with a single maximum, are called noninvertible, since, given x_{n+1}, one *cannot* solve $x_{n+1} = f(x_n)$ for x_n. This is due to the fact that for every value of x_{n+1} there are two possible values of x_n and therefore the functional relation cannot be inverted. If the index n is interpreted as the discrete time variable then the recursion relation generates new values of x_n forward in time but not backward in time; see eg. Ott (1981). This assumption corresponds to the reasonable requirement that the dynamic law stimulates x to grow when it is near zero, but inhibits its growth when it reaches some maximal value. An example of this is provided by the discrete version of the Verhulst equation for population growth that we have just examined

$$x_{n+1} = \mu \, x_n (1 - x_n/\Theta) \quad , \tag{4.1.22}$$

where you will recall that μ is the growth rate and Θ is the saturation level of the population.

Equation (4.1.22) is often called the discrete logistic equation and has been extensively studied in the physical sciences, usually in the scaled form with $y_n = x_n/\Theta$:

$$y_{n+1} = \mu \, y_n(1 - y_n) \quad .$$

(4.1.23)

Thus the scaling function is $f(y_n) = \mu y_n(1 - y_n)$ and when graphed vs. y_n yields the quadratic curve depicted in Figure (4.1.7).

The mapping operation is one that is accomplished by applying the function f to a given initial value y_0 to generate the successive images of this point. The point y_n is generated by applying the mapping f, n times to the initial point y_0:

$$y_n = f^n(y_0)$$

(4.1.24)

using the relation $f^n(y_0) = f(f^{n-1}(y_0))$. This is done graphically in Figure (4.1.7a) for $n=3$ using the rule: starting from the initial point y_0 a line is drawn to the function yielding the value $y_1 = f(y_o)$, then from the function to the diagonal (45 deg) line back to the function as shown to yield $y_2 = f(y_1) = f(f(y_0)) = f^2(y_0)$; etc. The intersection of the diagonal with the function f defines a point y^* having the property

$$y^* = f(y^*)$$

(4.1.25)

which is called a *fixed point* of the dynamic equation, i.e., y^* is the y_{ss} from (4.1.17). The fixed point corresponds to the equilibrium solution of the discrete equation and for (4.1.23) $y^* = 1 - 1/\mu$ (nontrivial) $y^* = 0$ (trivial). We can see in Figure (4.1.7b) that the iterated points are approaching y^* and as $n \to \infty$ they will reach this fixed point. To determine if a mapping will approach a fixed point asymptotically, i.e., if the fixed point is stable, we examine the slope of the function at the fixed point, see e.g., May (1972), Li and Yorke (1975) and Collet and Eckmann (1980). The function acts like a curved mirror either focusing the ray towards the fixed point under multiple reflections or diverging the ray away. The asymptotic directon (either towards or away from the fixed point) is determined by the slope of the function at y^*, which is depicted in Figure (4.1.8) by the dashed line and denoted by $f'(y^*)$ i.e., the (tangent) derivative of $f(y)$ at $y=y^*$. As long as $|f'(y^*)| < 1$ the iterations of the map are attracted to $y=y^*$. Again using the logistic map as an example we have $f'(y^*) = 2 - \mu$, so that the equilibrium point is stable and attracts all trajectories originating in the interval $0 < y < 1$ if and only if $1 < \mu < 3$.

When the slope of f is such that the fixed point becomes unstable, i.e., $|f'(y^*)| > 1$, then the solution "spirals" out. If the parameter μ is continuously increased until this instability is reached then the orbit will spiral out until it encounters a situation where $y_2^* = f(y_1^*)$ and $y_1^* = f(y_2^*)$, i.e., the orbit becomes periodic. Said differently, the mapping f has a *periodic orbit* of period 2 since $y_2^* = f(y_1^*) = f^2(y_2^*)$ and $y_1^* = f(y_1^*) = f^2(y_1^*)$

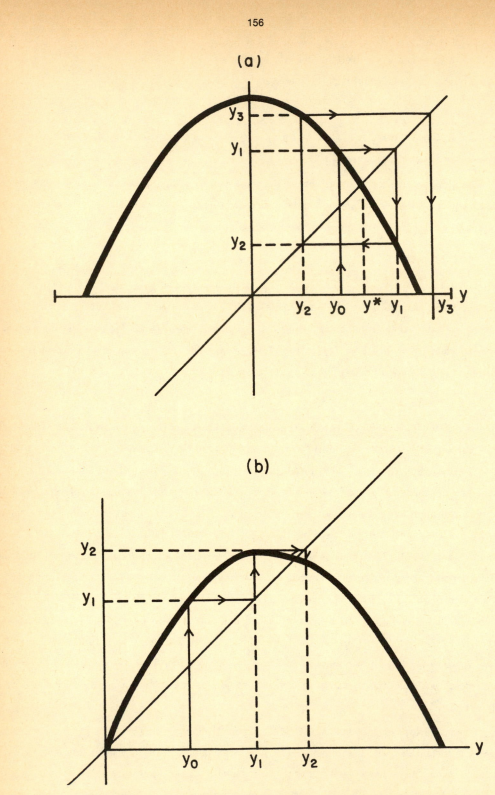

Figure 4.1.7. A mapping function with a single maximum is shown. In (a), the iteration away from the initial point y_o is depicted. In (b), the convergence to the stationary point y^* is shown.

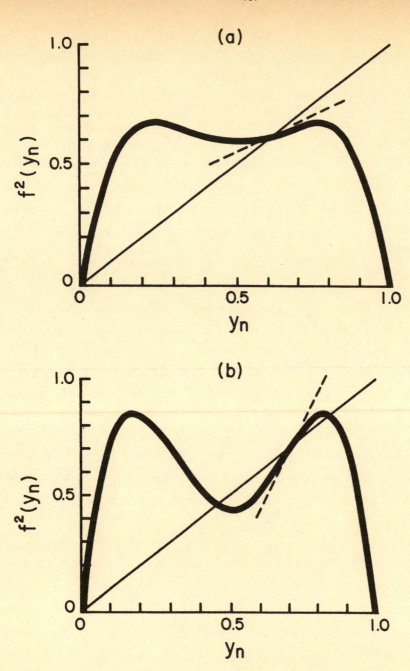

Figure 4.1.8.   The map f with a single maximum in Figure 4.1.7 yields an f^2 map with a double maximum. The slope at the point y^* is indicated by the dashed line and is seen to increase as the parameter μ is raised in the map from (a) to (b).

since y_1^* and y_2^* are fixed points of the mapping f^2 and not of the mapping f. In Figure (4.1.8a) we illustrate the mapping f^2 and observe it to have two maxima rather than the single one of f. As the parameter μ is increased further the dimple between the two maxima increases as do the height of the peaks along with the slopes of the intersection of f^2 with the diagonal; see Figure (4.1.8b).

For $1 < \mu < 3$ the fixed point is stable and y^* is a degenerate fixed point of f^2, ie, $y^* = f^2(y^*)$. At $\mu = 3.414$ the fixed point becomes unstable and two new solutions to the quadratic mapping emerge. These are the two intersections of the qadratic map with the diagonal having slopes with magnitude less than unity, y_1^* and y_2^*. The chain rule of differentiation of the derivative of f^2 at y_1^* and y_2^* is the product of the derivatives along the periodic orbit

$$f^{2\prime}(y_1^*) = f'(f(y_1^*))f'(y_1^*) = f'(y_1^*)f'(y_2^*) = f^{2\prime}(y_2^*) \tag{4.1.26}$$

so that the slope is the same at both points of the period 2 orbit, see eg. Li and Yorke (1975), and in fact the slope is the same at all k of the values of a period k orbit. This is in fact a continous process starting from the stable fixed point y^* when $|f'| < 1$; as μ is increased this point becomes unstable at $|f'| = 1$ and generates two new stable points with $|f^{2\prime}| < 1$ for a period 2 orbit; as μ is increased further these points become unstable at $|f^{2\prime}| = 1$ and generates four new stable points with $|f^{4\prime}| < 1$ for a period 4 orbit. This bifurcation sequence is tied to the value of the parameter μ. As this parameter is increased the discrete equation undergoes a sequence of bifurcations from the fixed point to stable cycles with periods 2, 4, 8, 16, 32, ... 2^k. In each case the bifurcation process is the same as that for the transition from the stable fixed point to the stable period 2 orbit. A graph indicating the location of the stable values of y for a given μ is given in Figure (4.1.9). Here we see that the μ interval between successive bifurcations is diminishing so that the "window" of values of μ wherein any one cycle is stable progessively diminishes. If we denote by μ_k the value of μ where the orbit bifurcates from length 2^{k-1} to 2^k, then

$$\lim_{k \to \infty} \frac{\mu_k - \mu_{k-1}}{\mu_{k+1} - \mu_k} = \textit{universal constant} \tag{4.1.27}$$

a result first obtained numerically by Feigenbaum (1978). This result indicates that a constant μ_∞ is being approach by this sequence. This critical parameter value is a point of accumulation of the period 2^k cycles. For equation (4.1.23) the critical value of this parameter is $\mu_\infty = 3.5700......$ The numerical value of μ_∞ is dependent on the particular map considered, although the existence of an accumulation point does not, and more importantly the universal constant in (4.1.27) has a value $4.69201 \ldots$ and is also independent of the specific choice of the map.

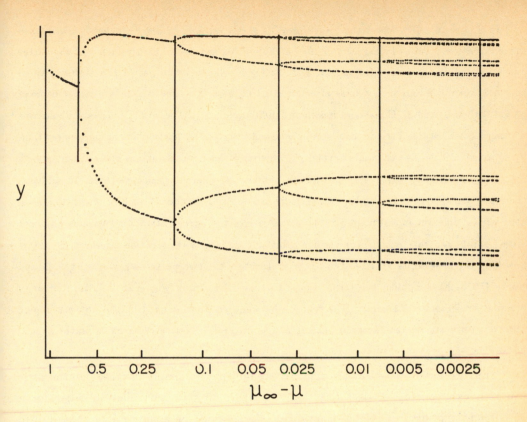

Figure 4.1.9. The bifurcation of the solution to the mapping $x \rightarrow 1 - \mu x^2$ as a function of $\mu_\infty - \mu$ is indicated. The logarithmic scale was chosen to clearly depict the bifurcation regions.

In Figure (4.1.9) we used the logarithm of μ as the absissa in order to clearly distinguish the bifurcation points. In Figure (4.1.10) we replot this sequence linearly in μ. In the latter figure we distinguish from left to right, a stable fixed point, orbit of period 1; a stable orbit of period 2, then 4, 8 and then a haze of orbits starting along the line μ_∞, then another orbit of period 6 then 5, and 3. Collet and Eckmann (1980) comment: "The astonishing fact about this arrangement of stable periodic orbits is its *independence* of the particular one-parameter family of maps." The haze of points beyond μ_∞ consists of an infinite number of fixed points with different periodicities, along with an infinite number of different periodic orbits. In addition there are an uncountable number of aperiodic trajectories (bounded) each of which is associated with a different initial point y_o. As in the case of the Lorenz attractor two such adjacent initial points generate orbits that become arbitiarly distant with iteration number; no matter how long the time series generated by $f(y)$ is iterated, the two patterns never repeat. Li and Yorke (1975) have applied the term *chaotic* to this hazy region where an infinite number of different trajectories can occur. The relation between *chaos* as defined by the above mapping and that on a strange attractor is unclear and is an area of active research in mathematical physics as well as in mathematics.

Thus we have arrived at the remarkable fact that a simple discrete deterministic equation can generate trajectories that are aperiodic. In particular in order for a one-dimensional map to exhibit chaotic behavior, it must be noninvertible. May (1976) points out a number of practical implications of this result. The first being

". . . that the apparently random fluctuations in census data for an animal population need not necessarily betoken either the vagaries of an unpredictable environment or sampling errors: they may simply derive from a rigidly deterministic population growth relationship such as equation (4.1.1)."

We emphasize that the chaos encountered in these one-dimensional maps is quite distinct from that found by Lorenz for his continuous dissipative system. Said differently, the chaos found on a strange attractor and that found using a one-dimensional noninvertible map may not be the same. In the next section we discuss the discrete analog of the Lorenz attractor.

4.1.3 More Discrete Systems (New but Related Chaos)

In the preceding section we defined a mapping in terms of a projection of a higher order dynamic system onto a one-dimensional line. This same definition can be applied for the intersection of the trajectories of a higher order dynamic process with a two-dimensional plane. In Figure (4.1.11) a sketch of a trajectory in three dimensions is shown, the intersection of the orbit with the plane in one direction defines a set of points that can be obtained by means of the two-dimensional map:

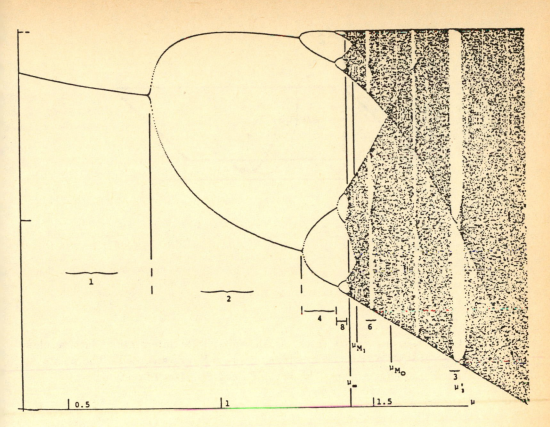

Figure 4.1.10.   The same as Figure 4.1.9, but with a linear scale in $\mu_\infty - \mu$ so that the hazy region denoting chaos is clearly observed.

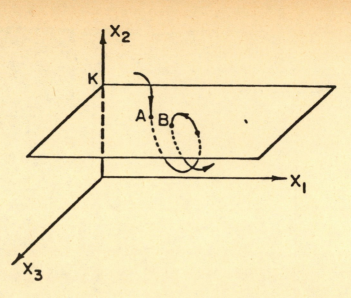

Figure 4.1.11.   An arbitrary trajectory is shown and its intersection with a plan parallel to the $x_1 - x_3$ - plane at $x_2 = K$ are recorded. The points A, B, ... define a map. This is the Poincaré surface of section.

$$x_{n+1} = f_1(x_n, y_n) \qquad y_{n+1} = f_2(x_n, y_n) \ . \qquad (4.1.28)$$

Here we follow Ott (1981) and consider only invertible maps where (4.1.28) can be solved uniquely for x_n and y_n as functions of x_{n+1} and y_{n+1}; $x_n = g_1(x_{n+1}, y_{n+1})$ and $y_n = g_2(x_{n+1}, y_{n+1})$. If n is the time index then invertibility is equivalent to time reversibility, so that these maps are reversible in time whereas those in the preceding section were not. The maps in this section are analogous to the dynamic equations discussed in physics and chemistry.

The reason for examining higher order maps, such as the two-dimensional example given by (4.1.28), is that under certain conditions these maps share many of the properties of the Lorenz attractor. Thus we may anticipate that complex systems in the biological and behavioral sciences for which discrete equations may be a more natural way to model the dynamics than would be the traditional continuous equations of the physical science, do not have to be reduced to one-dimensional maps in order to see chaos emerge. In particular we will establish the connection between these invertible maps and the strange attractor of Lorenz as well as the fractal dimension discussed in the context of the Lévy distribution [cf. §3.2].

The one-dimensional noninvertible maps of the preceding section were obtained by projecting a higher order trajectory onto a one-dimensional line. Let us now reverse the process and expand the space of the noninvertible map from one to two-dimensions by introducing the coordinate y_n in the following way:

$$x_{n+1} = f(x_n) + y_n \qquad (4.1.29a)$$

$$y_{n+1} = \beta x_n \ . \qquad (4.1.29b)$$

Of course, f is noninvertible and $\beta = 0$ collapses (4.1.29) back onto the one-dimensional map (4.1.9). For any non-zero β, however, the map (4.1.29) is invertible, i.e., $x_n = y_{n+1}/\beta$ and $y_n = x_{n+1} - f(y_{n+1}/\beta)$. Thus we have transformed a noninvertible map to an invertible one by extending the space. As Ott (1981) points out, however, if β is sufficiently small the distinction between the invertible two-dimensional map and the noninvertible one-dimensional map may not be measurable.

For the Lorenz attractor we examined the behavior of a small phase space volume obtained by perturbing the solutions of the equations of motion and found that the volume contracts due to dissipation. The analogous quantity associated with the mapping (4.1.29) is the Jacobian of the map:

$$J \equiv \begin{vmatrix} \dfrac{\partial x_{n+1}}{\partial x_n} & \dfrac{\partial x_{n+1}}{\partial y_n} \\[2mm] \dfrac{\partial y_{n+1}}{\partial x_n} & \dfrac{\partial y_{n+1}}{\partial y_n} \end{vmatrix} = -\beta \qquad (4.1.30)$$

which is obtained by using (4.1.29) directly in the definition of the Jacobian. Thus the volume at consecutive times is given by

$$V_{n+1} = -\beta V_n \qquad (4.1.31)$$

which for an initial volume V_o has the solution

$$V_{n+1} = (-1)^{n+1} \beta^{n+1} V_o , \qquad (4.1.32)$$

so that if $|\beta| < 1$ the volume will contract by a factor $|\beta|$ at each application of the mapping just as does the volume in the continuum case. We now know that this contraction does *not* imply that the solution goes over to a point in phase space, but only that it is attracted to some bounded region of dimension lower than that of the initial phase space. If the dimension of the attractor is non-integer, then the attractor is fractal; see eg. in Mandelbrot (1980) where the observation that the fractal dimension of a set may or may not be consistent with the term *strange* as used earlier. Following Eckmann (1981), we employ the property that, if all the points in the initial volume V_0 converge to a single attractor, but that points which are arbitrarily close initially separate exponentially in time, then that attractor is called strange. This property of nearby trajectories to exponentially separating in time is called *sensitive dependence on initial conditions* and gives rise to the aperiodic behavior of strange attractors. There exists however a large variety of attractors which are neither periodic orbits nor fixed points and which are not strange attractors. All of these, states Eckmann (1981), seem to present more or less pronounced chaotic features. Thus there are attractors that are erratic but not strange. We will not pursue this general class here.

As an example of the two-dimensional invertible mapping we first transform the logistic equation (4.1.22) into the family of maps $x_{n+1} = 1 - cx_n^2$ with the parametric identification $c = (\mu/2-1)\mu/2$ and $0 < c \le 2$, since $2 < \mu \le 4$ and x_n maps the interval $[-1, 1]$ onto itself. Then using (4.1.29) we obtain the mapping first studied by Henon (1976):

$$x_{n+1} = 1 - cx_n^2 + y_n \qquad (4.1.33a)$$

$$y_{n+1} = \beta x_n \qquad (4.1.33b)$$

In Figure (4.1.12) we have copied the loci of points for the Henon system (4.1.33) in which 10^4 successive points from the mapping with the parameter values $c = 1.4$ and $\beta = 0.3$ initiated from a variety of choice of (x_0, y_0). Ott points out that, as the map is iterated, points come closer and closer to the attractor eventually becoming indistinguishable from it. This is however an illusion of scale. If the boxed-in

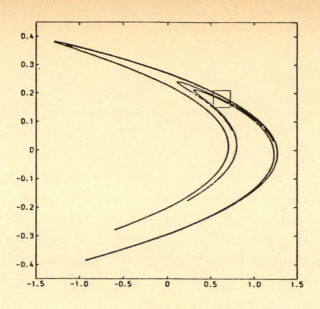

Figure 4.1.12.   Iterated point of the map (4.1.21), for 10^4 iterations [from Ott (1982)].

region of the figure is magnified one obtains Figure (4.1.13a) from which a great deal of structure of the attractor can be discerned. If the boxed region in this latter figure is magnified, then what had appeared as three unequally spaced lines appear in Figure (4.1.13b) as three distinct parallel intervals containing structure. Notice that the region in the box of Figure (4.1.13a) appears the same as that in (4.1.13b). Magnifying the boxed region in this latter figure we obtain Figure (4.1.13c), which aside from resolution is a self-similar representation of the structure seen on the two preceding scales. Thus we observe scale invariant, Cantor-set-like structure transverse to the linear structure of the attractor. Ott (1981) concludes that because of this self-similar structure the attractor is probably strange as is the Lorenz attractor. In fact it has been verified by direct calculation that initially nearby points separate exponentially in time; see Feit (1978), Curry (1979) and Simo (1979), thereby coinciding with our definition of the strange attractor.

4.2 Growth, Competition and Avoidance

In an earlier section (§3.3) we discussed the construction of models to describe the growth of populations and emphasized the importance of nonlinear terms that lead to the asymptotic saturation of the population level. Such descriptions excluded from explicit consideration the effect of competition among different species as a mechanism for limiting the growth of a given population. Such interactions can be very important, however, especially when one species is the food source for another, as we show below. The attentive reader may object to this distinction and point out that the model of technological replacement considered in §3.3 does in fact involve competition over a fixed quantity, i.e., the percent of the market captured by a product. That person would be correct, however the notion of competition as a behavioral mechanism does not appear in the earlier discussion. A second concept which was mentioned but not significantly developed was that of spatial mobility as a way of maintaining population growth without over-burdening the local habitat's capacity to support a population. Certain of these ideas are developed below.

4.2.1 Prey-Predator Systems

One of the first successes of mathematical ecology was the demonstration that certain interacting species undergo periodic variations in population level. Ecology is the science of the interaction among various plants and animals and its first mathematical formulation was made by Lotka (1925). Volterra (1931) subsequently contributed to the development and solution of a model of a community in which organisms of one population provide food for those of the other. Such bio-systems are referred to as *prey-predator* systems. However, the periodicities arising from the

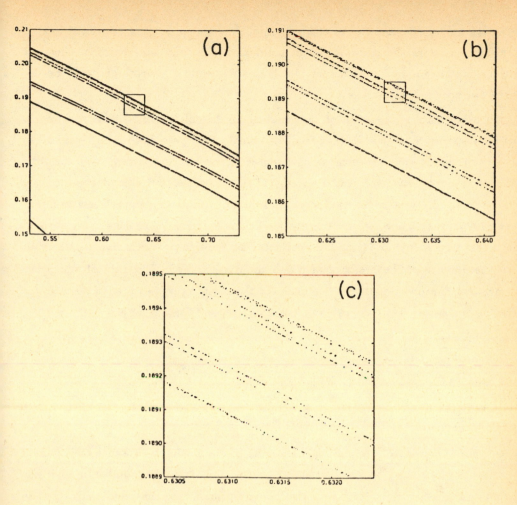

Figure 4.1.13: (a) Enlargement of the boxed region in Figure 4.1.12, 10^5 iterations; (b) enlargement of the square in (a), 10^6 iterations; (c) enlargement of the square in (b), 5×10^6 iterations [from Ott (1982)].

interaction among the constituent elements in such a system are not restricted to ecological systems, e.g., the rise and fall of political parties and the chains of chemical reactions can have a similar structure. In this section we concentrate primarily on the form of two-species interactions.

We begin with a discussion of the classical Lotka-Volterra model. A species S_2 feeds on a species S_1 which, in turn feeds on some source present in such large excess in the environment that the mass of this source is considered constant over the time period of interest. Montroll and Badger (1974) review the model and consider the species S_1 and S_2 to be small fish (prey) and big fish (predator), respectively, whereas Lotka (1925) was somewhat more general. The population of species S_1 (prey) at time t is denoted by $N_1(t)$ and is postulated to grow in a Malthusian way at rate k_1, in the absence of the predator of which there are $N_2(t)$. The rate at which the prey are lost is proportional to the number of times the prey and predator encounter each other pairwise. Thus the rate of change in the prey population is

$$\frac{dN_1(t)}{dt} = k_1 N_1(t) - \lambda_1 N_1(t) N_2(t) \tag{4.2.1}$$

with a related expression for the predator population

$$\frac{dN_2(t)}{dt} = -k_2 N_2(t) + \lambda_2 N_1(t) N_2(t) \ . \tag{4.2.2}$$

We see that the predator population is assumed to die out in the absence of prey at a rate given by k_2 and to grow at a rate proportional to the number of pairwise encounters with the prey. Note that the loss constant λ_1 for the prey is not equal to the gain constant λ_2 for the predator, since these constants indicate the efficiency with which the encounters delete the prey population and are converted into predator.

For convenience we define the normalized populations

$$\eta_1(t) = \lambda_2 N_1(t)/k_2 \quad \text{and} \quad \eta_2(t) = \lambda_1 N_2(t)/k_1 \tag{4.2.3}$$

and substituting these quantities into (4.2.1) and (4.2.2) obtain

$$\dot{\eta}_1 = k_1 \eta_1 (1 - \eta_2) \tag{4.2.4a}$$

$$\dot{\eta}_2 = -k_2 \eta_2 (1 - \eta_1) \tag{4.2.4b}$$

so that the ratio of these two equations yields

$$\frac{d\eta_1}{k_1 \eta_1 (1 - \eta_2)} = \frac{d\eta_2}{-k_2 \eta_2 (1 - \eta_1)} \ . \tag{4.2.5}$$

Cross multiplying we obtain

$$\frac{d\eta_1(1-\eta_1)}{k_1\eta_1} = \frac{d\eta_2(1-\eta_2)}{-k_2\eta_2} = \text{constant} \tag{4.2.6}$$

which immediately integrates to

$$(\eta_1 \, e^{-\eta_1})^{1/k_1} \, (\eta_2 \, e^{-\eta_2})^{1/k_2} = constant \tag{4.2.7}$$

the result first obtained by Volterra (1931). Equation (4.2.7) relates the dynamic behavior of the prey population $(\eta_1(t))$ to that of the predator population $\eta_2(t)$).

Introducing the transformations $U \equiv (\eta_1 \, e^{-\eta_1})^{1/k_1}$ and $V \equiv (\eta_2 e^{-\eta_2})^{1/k_2}$ (4.2.7) reduces to the equation for a hyperbola; $UV = \text{constant}$. The plot of this hyperbola is shown in Figure (4.2.1a). In Figures (4.2.1b) and (4.2.1c) the $V(\eta_2)$ and $U(\eta_1)$ are shown to have maxima denoted by M_1 and M_2 in these figures. The dependence of the solution on η_1 and η_2 is depicted in Figure (4.2.1d) where the maxima constrain the solution to lie in a confined region of phase space whose shape depends on the initial conditions. Hence the relevant region of the hyperbola in Figure (4.2.1a) is bounded by the points A and B. The points in this interval are mapped onto the limit cycle in Figure (4.2.1d).

In Figure (4.2.2) the solutions $\eta_1(t)$ and $\eta_2(t)$ of (4.2.4) are graphed as a function of time for a variety of initial conditions and values of the parameters k_1 and k_2. These curves are taken from Montroll and Badger (1974) and are seen to vary sinusoidally in time. These smooth theoretical trajectories are to be contrasted with the dynamics of the lynx and hare population in Canada for the period 1845-1935 according to the data of the Hudson Bay Company [taken from MacLuich (1937)]; see Figure (4.2.3). The deviations of the data from the prediction of the Lotka-Volterra model suggest that the coupling of these species to the environment may be both non-negligible and random. Additional complications of this type are considered, for example, by Goel et al. (1971), Montroll and Badger (1974), Svirezhev and Logofet (1983), among others.

There is a fundamental difficulty with the classical Lotka-Volterra model as was pointed out by May (1972), which is that the equations (4.2.1) and (4.2.2) are neutrally stable. The populations of the prey and predator will oscillate, as observed for systems in nature, but the amplitudes of these oscillations will depend on how the state was initiated. This situation was known to Lotka who had at first claimed that the solutions to his equations lay on a limit cycle, see Lotka (1920), but later in his 1925 book acknowledged his error and commented that his solutions were dependent on the initial state of the system. Neutral stability is unacceptable because it means that if the system is disturbed it will move to an adjacent orbit rather than symptotically

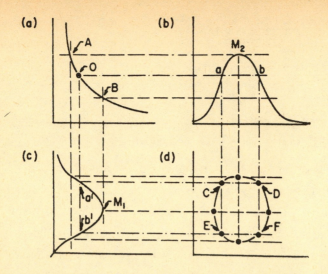

Figure 4.2.1.   The phase space projections of the periodic solutions to the Lotka-Volterra equations. The orbit looks like a limit cycle, but is in fact a neutrally stable, initial condition dependent solution.

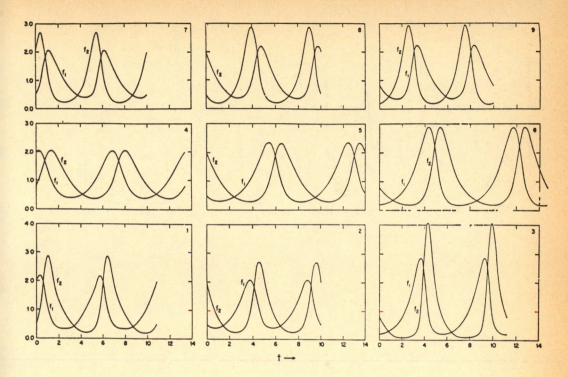

Figure 4.2.2.   Time variations of two populations for several values of parameters, (1), (2), (3), $k_1 = 1$, $k_2 = 2$; (4), (5), (6), $k_1 = k_2 = 1$; (7) (8) (9), $k_1 = 2$, $k_2 = 1$ [from Montroll and Badger (1974)].

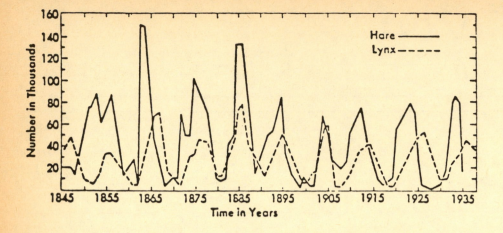

Figure 4.2.3. Changes in the abundance of lynx and snowshoe hare
[from D.A. MacLuich (1937)].

returning to the orbit on which it started out. Thus, as discussed by Svirezhev and Logofet (1983) these orbits lack *roughness* or as we would say the solutions are not *robust*. May (1972) makes the following closing comment:

" .... to regard the familiar and regular oscillations of the lynx and hare populations recorded by the Hudson's Bay Trading Company as resulting from a pure Lotka-Volterra oscillation about a neutrally stable equilibrium point, which is to say, having an amplitude determined by some environmental shock over 100 years ago, is quite implausible; this system, with maximum hare population being constant to within a factor of 2 over 100 years or nine cycles, is surely the outcome of some stable limit cycle."

Let us consider a more general model of the prey-predator system by introducing the arbitrary functions F_1 and F_2:

$$\frac{dN_1}{dt} = N_1 F_1(N_1, N_2) \qquad (4.2.8a)$$

$$\frac{dN_2}{dt} = N_2 F_2(N_1, N_2) \qquad (4.2.8b)$$

The functions F_1 and F_2 are arbitrary as to their specific functional forms but not as to the general properties they must have in order to represent a reasonable interactive system. For example, in the absence of predators the prey population cannot grow indefinitely as assumed in (4.2.1), but as discussed by Verhulst [cf. § 3.2] the limited resources will induce a mechanism for saturation in the dynamic equations for the population. Therefore the function $F_1(N_1, N_2)$ must have a saturation inducing part, say $F_1(N_1, 0) = k_1[1 - N_1/\Theta_1]$. On the other hand, (4.2.2) assumes that the amount of prey consumed by a predator in a unit time can increase indefinitely with increasing prey population. This assumption violates certain physiological constraints, see e.g., §3.3. May (1972) gives a typical example from the ecology literature for replacing the Lotka-Volterra equation set by

$$\frac{dN_1}{dt} = k_1 N_1 [1 - N_1/\Theta_1] - \lambda_1 N_2 [1 - e^{-\beta_1 N_1}] \qquad (4.2.9a)$$

$$\frac{dN_2}{dt} = -k_2 N_2 + \lambda_2 N_2 \left[1 - e^{-\beta_2 N_1}\right] \qquad (4.2.9b)$$

where for low levels of the population the exponential can be expanded to again yield the binary terms, but in general there is a saturation to the prey-predator interaction. The rate constants and other parameters in this example are not important for the present disscussion, except to note that (4.2.9) yields solutions that are limit cycles and are therefore stable against perturbations. Kolmogorov (1936) in fact constructed a theorem that tells us that such models posses either a stable equilibrium point or a stable limit cycle. He first proved this theorem in a general prey-predator context, e.g., (4.2.8); it is discussed at length by Svirezhev and Logofet (1983), although the condensed discussion of May (1972) is very readable.

A few comments may also be in order regarding competition as a general mechanism operative in social behavior; as for example in the spread of ideas and rumors through a population, the replacement of new technology by old [cf. §3.3], the spread of cultural traits, of fads and fashions, and of opinion, to name a few. Coleman (1964) discusses such processes from the point of view of social theorists who have made diffusion their central mechanism of social change. He points out that the most sophisticated mathematical work and the most serious empirical studies, up until that time, had been in medical epidemics. That observation remains true today. The basic model for describing the contagion of a disease throughout a population assumes complete intermixing of the population with the rate of propagation being proportional to the product of the number of infectives $N_1(t)$ and the number of susceptibles $N_2(t)$. Thus the deterministic form of the classical social diffusion is

$$\frac{dN_1}{dt} = kN_1N_2 \tag{4.2.10}$$

where k is the rate of conversion from the *haves* N_1 to the *have—nots* N_2. This equation gives rise to the familiar logistic curves for N_1 as a function of time since the total population N is a constant $N = N_1 + N_2$, so that the right hand side of (4.2.10) can be written $kN_1(N - N_1)$.

Equation (4.2.10) describes a population where each individual is assumed to have equal numbers of contacts with each other individual, since only in that case is N_1N_2 equal to the number of relations that exist between the two groups. In realistic situations those individuals which are near each other in space interact more frequently than those that are separated by some distance. Thus the populations should be functions of the spatial coordinate x and the *physical diffusion* of infection is modeled in one spatial dimension as follows:

$$\partial_t N_1(x,t) = k N_1(x,t) N_2(x,t) + D \partial_{xx} N_1(x,t) \tag{4.2.11}$$

or in terms of the normalized population $\rho(x,t) = N_1(x,t)/N$ we have

$$\partial_t \rho = k\rho(x,t) [1 - \rho(x,t)] + D \partial_{xx} \rho(x,t) . \tag{4.2.12}$$

R.A. Fisher (1937) constructed (4.2.12) to discuss the problem of the propagation of a virile mutant in a population in a one-dimensional habitat, i.e., along the coast of an island. Here $\rho(x,t)$ is the fraction of the population to the value N, the saturation population, per unit length. At the same time Kolmogorov, Petrovsky and Piscounoff (1937) investigated a general class of partial differential equations which describe simultaneous growth and diffusion processes. In two spatial dimensions they considered the general class of equations

$$\partial_t \rho = D\left[\partial_{xx} + \partial_{yy}\right]\rho + F(\rho) \tag{4.2.13}$$

where $F(\rho)$ is a saturation inducing growth function for the population fraction $\rho(x,y,t)$ and of which (4.2.12) is a special case.

The direct solution of Fisher's equation (4.2.12) is extremely difficult to obtain due to the nonlinear structure of the equation. Fisher (1937) ahd Skellam (1954) obtained numerical solutions assuming a solution of the form $\rho(x,t) = \rho(x-vt)$, i.e., a diffusion wave travelling at a speed v. A more general analytic solution may be obtained in a restricted region if we consider the expansion

$$\rho = \exp\{\ln\rho\} \simeq 1 + \ln\rho + o(\ln\rho^2) \tag{4.2.14}$$

for the population in spatial regions near saturation, i.e., $\rho \approx 1$. Substituting this expansion in (4.2.12) yields the new equation

$$\partial_t \rho(x,t) = D\left\{\partial_{xx}\rho(x,t) - \frac{1}{\rho(x,t)}\left[\partial_x \rho(x,t)\right]^2\right\} - k\rho(x,t)\ln\rho(x,t) \tag{4.2.15}$$

which has the more general form of the Kolmogorov, Petrovsky and Piscounoff equation (4.2.13) with the proper interpretation of $F(\rho)$. Although (4.2.15) looks even more formidable than the Fisher equation, it is in fact quite friendly. The change of variables $u(x,t) = e^{-kt}\ln\rho(x,t)$ leads to the simple diffusion equation

$$\partial_t u(x,t) = D\,\partial_{xx} u(x,t) \tag{4.2.16}$$

discussed in §2.3. For the initial distribution $u(x,0)$ the diffusion equation has the solution

$$u(x,t) = \int_{-\infty}^{\infty} \frac{e^{-\frac{(x-x')^2}{4Dt}}}{\sqrt{2\pi Dt}}\, u(x',0)dx' \tag{4.2.17}$$

or in terms of the population variable

$$\rho(x,t) = \exp\left\{\frac{e^{-kt}}{\sqrt{2\pi Dt}} \int_{-\infty}^{\infty} \exp\left[-\frac{(x'-x)^2}{4Dt}\right]\ln\rho(x',0)\, dx'\right\}. \tag{4.2.18}$$

Thus (4.2.18) is the analytic solution to (2.4.15) for an initial distribution $\rho(x,0)$ and is therefore the solution to Fisher's equation in a region near saturation.

In a correspondingly approximate way we may seek a solution to (4.2.12) in a region quite far from saturation, i.e., where $\rho \ll 1$, so that

$$\partial_t \rho(x,t) = D \partial_{xx} \rho(x,t) + k\rho(x,t) \quad . \tag{4.2.19}$$

The solution to (2.4.19) is found to be

$$\rho(x,t) = \frac{e^{kt}}{\sqrt{2\pi Dt}} \int_{-\infty}^{\infty} \exp\left[-\frac{(x-x')^2}{4Dt} \right] \rho(x',0)dx' \tag{4.2.20}$$

which clearly diverges in time. This divergence is vitiated by matching the two asymptotic solutions at an intermediate time. Thus as long as the true behavior of the population is monotonic the two extreme solutions should joint smoothly together.

4.2.2 Interacting Random Walkers

Let us consider the system composed of two interacting species from a totally different point of view. The process of diffusion in such a system generally tends to randomize the components. In a molecular system the intermolecular forces are often such that aggregation is the preferred state. Social forces within ethnic groups also appear to have a similar aggregating effect (see discussion in §3.2). In this section we shall be concerned with a nonlinear random walk model which can be used to describe nonlinear rate processes in macroscopic population systems characterized by this competition between diffusion and the forces acting between members of the population.

Consider a single species of random walkers performing a one-dimensional walk in a manner in which each step is a function of the concentration of walkers to the left and right of the one being examined. Let our linear space be divided into cells, each of length a. Let ρ be the average density of walkers (per unit length) averaged over all the cells and $N=\rho a$ be the mean number of walkers in a given cell. At any time the number of walkers in a given cell may be more or less than this number. We follow the motion of a single walker who, we suppose, takes steps at regular time intervals with pauses of duration τ. We postulate that after his νth step (at time $\nu\tau$) he prepares to take a step from cell k (at position ka from the origin) in a manner biased by the local concentration of walkers in cells $(\nu-1)$ and $(\nu+1)$. We define $N(ka,\nu\tau)$ as the number of walkers in the kth cell at time $\nu\tau$ after the walk beings $(\nu = 0)$. The above bias can be written in terms of the local concentration

$$\rho(k,\nu) = \frac{N(ka,\nu\tau)}{N} \tag{4.2.21}$$

as

$$p(k \to k+1) = \frac{1}{2}\left\{ 1+\mu\left[\rho(k+1,\nu) - \rho(k-1,\nu)\right] g\left[\rho(k,\nu)\right] \right\} \tag{4.2.22}$$

and

$$q(k \rightarrow k-1) = 1 - p(k \rightarrow k+1)$$

$$= \frac{1}{2} \left\{ 1 + \mu \left[\rho(k-1,\nu) - \rho(k+1,\nu) \right] g \left[\rho(k,\nu) \right] \right\} . \qquad (4.2.23)$$

The probability of a step from k to $k+1$ is given by Equation (4.2.22) and the probability of a step from k to $k-1$ is given by Equation (4.2.23); $\rho(k,\nu)$ is the population density (concentration) on cell k at time $\nu \tau$, assuming the random walk commenced at $\nu = 0$.

The qualitative features of the transition probabilities p and q are easily analyzed. Consider the case were $g(\rho)$ is positive definite:

(a) If the concentration of walkers at site $k+1$ exceeds that at $k-1$, then the transition of a walker site k to higher k is favored. When the inequility is reversed, transitions to lower k are favored.

(b) As the coupling parameter μ in (4.2.22) and (4.2.23) goes to zero, or as the concentration at site $k+1$ approaches that at $k-1$, the bias vanishes, resulting in an unbiased random walk, ie., equal probability of a step either right or left.

(c) We have as yet left the weighting function $g(\rho)$ arbitrary. If the function is set equal to a constant, unity say, then the bias depends only on the difference between the concentration between the $k+1$ and $k-1$ sites, i.e., the concentration gradient. This dependence would be the same if the local concentration $\rho(k,\nu)$ were 0.1 of 0.9. One might expect, however, that in some cases the bias might be less when the local concentration is 0.9 than when it is 0.1. This dependence is reflected in the choice of the weighting function, for example if $g[\rho(k,\nu)]$ had the form

$$g\left[\rho(k,\nu) \right] = \frac{1 - \rho^{\alpha}(k,\nu)}{\alpha} , \qquad (4.2.24)$$

it would reflect this latter behavior. The decision on the selection of the form of the weighting function may be postponed until later. We will see that the solution to the nonlinear equation which results from the above random walk is quite sensitive to the choice of the weighting function.

Using the probabilities of right and left transitions, we find that the concentration at site k at time $(\nu+1)\tau$ can be written in terms of the concentration at neighboring sites at time $\nu\tau$,

$$\rho(k,\nu+1) = \frac{1}{2}\,\rho(k-1,\nu)\left\{1+\mu[\rho(k,\nu)-\rho(k-2,\nu)]g\,[\rho(k-1,\nu)]\right\}$$

$$+\frac{1}{2}\,\rho(k+1,\nu)\left\{1+\mu\,[\rho(k,\nu)-\rho(k+2,\nu)]\,g\,[\rho(k+1,\nu)]\right\}\ . \tag{4.2.25}$$

We introduce the notation

$$\Delta\rho(k) \equiv \rho(k+1)-\rho(k) \tag{4.2.26}$$

into (4.2.25) and rewrite the expression as

$$\tau^{-1}[\rho(k,\nu+1)-\rho(k,\nu)] = \frac{1}{2}\,\frac{a^2}{\tau}\left[\frac{\Delta\rho(k,\nu)-\Delta\rho(k-1,\nu)}{a^2}\right]$$

$$+\frac{1}{2}\,\frac{\mu a}{\tau}\,[\Delta\rho(k-1,\nu)+\Delta\rho(k-2,\nu)]\,\rho(k-1,\nu)\,\frac{g\,[\rho(k-1,\nu)]}{a}$$

$$-\frac{1}{2}\,\frac{\mu a}{\tau}\,[\Delta\rho(k,\nu)+\Delta\rho(k+1,\nu)]\,\rho(k+1,\nu)\,\frac{g\,[\rho(k+1,\nu)]}{a} \tag{4.2.27}$$

If we let $\tau \rightarrow 0$, then the left side of Equation (4.2.27) tends to $\partial_t\rho$; if we also let

$$D = \frac{a^2}{2\tau} \quad \text{and} \quad \psi = \frac{a^2\mu}{\tau} \tag{4.2.28}$$

where D and ψ remain finite as $a \equiv dx \rightarrow 0$ and $\tau \equiv dt \rightarrow 0$ we find that,

$$\partial_t\rho = \partial_x(\mathrm{D}\,\partial_x\rho) \tag{4.2.29}$$

where

$$\mathrm{D} = D - 2\psi g\rho\ . \tag{4.2.30}$$

Equation (4.2.29) has the structure of a diffusion equation in an anisotropic medium and clearly indicates the dependence of the diffusion coefficient on the weighting function $g(\rho)$. From the postulated form of the random walk process the weighting function may be interpreted in terms of the interaction between the members of the given species. The form of the diffusion equation which results from the interacting random walkers is therefore not completely unexpected, since it is the interspecies interaction which leads to a nonconstant diffusion coefficient in general. The new result shows, however, how the diffusion coefficient and the weighting function are related.

Equation (4.2.29) conserves species number,

$$\int \partial_t\rho\ dx = 0 \tag{4.2.31}$$

which expresses the fact that the random walk process constitutes a closed system in which the total number of walkers is a constant. If we relax this restriction, i.e., let the total number of walkers vary, then our system is open to outside influence. Traditionally, one discusses this variation in the population in terms of birth and death processes, but we will find it more useful to talk of immigration to and emmigration from a given location.

The notion of immigration may be thought of as a birth-type of process in that the number of walkers at a point is increased by a given amount. The source of these walkers is from outside the system and they are attracted over a long range into the system. The notion of emmigration may in the same sense be thought of as a death-type process in which walkers are seduced out of the system by some long range collective force. The enhancement and depletion of the walker population are competitive mechanisms and may completely cancel one another. The inclusion of this process in the attractive random walk process will give us a general model of the interaction modes available to a large population.

We again construct our finite difference equation for the random walkers who interact with each other but now include the processes of immigration and emmigration:

$$I(k,\nu;\rho) = \frac{\epsilon}{2} [\rho(k+1,\nu) + \rho(k-1,\nu)] g [\rho(k,\nu)] \tag{4.2.32}$$

and

$$E(k,\nu;\rho) = \epsilon\rho(k,\nu) g [\rho(k,\nu)] \quad . \tag{4.2.33}$$

The qualitative features of these equtions (for $g(\rho) > 0$) are:

(a) Walkers outside the system are attracted by the average concentration of walkers in the neighborhood of site k by $I(k,\nu;\rho)$. In the continuous limit this average concentration approaches that at site k. The weighting function $g(\rho)$ may be used to enhance the immigration to sites where the average concentration is high if its functional form is properly chosen. This corresponds to decreasing the bias for high concentration and increasing the bias for low concentration.

(b) Walkers inside the system are not so much seduced away from the system as modeled by $E(k,\nu;\rho)$ but are rather repelled by the local concentration. When the local concentration is high there is an urge to emmigrate, but this is partially suppressed by the weighting function which attracts the walkers back. The degree of emmigration is therefore quite sensitive to the functional form of the interspecies interaction.

Using the probabilities of right and left transitions defined by (4.2.22) and (4.2.23) we write for the concentration at site k and time $(\nu + 1)\tau$

$$\rho(k,\nu + 1) = p(k-1 \rightarrow k)\rho(k-1,\nu) + q(k+1 \rightarrow k)\rho(k+1,\nu)$$

$$+ [I(k,\nu) - E(k,\nu)]\rho(k,\nu) \quad ,$$

or more explicitly,

$$\rho(k,\nu+1) \;=\; \frac{1}{2}\,\rho(k-1,\nu)\,\left\{1+\mu\,[\rho(k,\nu)-\rho(k-2,\nu)]\,g\,[\rho(k-1,\nu)]\right\}$$

$$+\frac{1}{2}\,\rho(k+1,\nu)\,\left\{1+\mu\,[\rho(k,\nu)-\rho(k+2,\nu)]\,g\,[\rho(k+1,\nu)]\right\}$$

$$+\frac{\epsilon}{2}\,\rho\,(k,\nu)\,[\rho(k+1,\nu)+\rho(k-1,\nu)-2\rho(k,\nu)]\;g\,[\rho(k,\nu)] \tag{4.2.34}$$

in terms of the concentrations at neighboring sites. Again we introduce the notation $\Delta\rho(k)$ and rewrite Equation (4.2.34) as,

$$\tau^{-1}\,[\rho(k,\nu+1)-\rho(k,\nu)] \;=\; \left\{\frac{1}{2}\,\frac{a^2}{\tau}+\frac{\epsilon a^2}{2\tau}\,\rho(k,\nu)g\,[\rho(k,\nu)]\right\}\,\frac{\Delta\rho(k,\nu)-\Delta\rho(k-1,\nu)}{a^2}$$

$$+\frac{1}{2}\,\frac{\mu a}{\tau}\,[\Delta\rho(k-1,\nu)+\Delta\rho(k-2,\nu)]\,\rho(k-1,\nu)\,\frac{g\,[\rho(k-1,\nu)]}{a}$$

$$-\frac{1}{2}\,\frac{\mu a}{\tau}\,[\Delta\rho(k\nu)+\Delta\rho(k+1,\nu)]\,\rho(k+1,\nu)\,\frac{g\,[\rho(k+1,\nu)]}{a} \tag{4.2.35}$$

We take the limit of this equation as $\tau\to 0$, letting the left side approach $\partial_t\rho$ and introduce

$$D \;=\; \frac{a^2}{2\tau}\;,\qquad \psi \;=\; \frac{a^2\mu}{\tau}\quad\text{and}\quad \phi \;=\; \frac{\epsilon a^2}{2\tau} \tag{4.2.36}$$

where D, ψ and ϕ remain finite as $a\equiv dx\to 0$ and $\tau\equiv dt\to 0$ and obtain for (4.2.35)

$$\partial_t\rho \;=\; [D+(\phi-2\psi)\,g\rho]\,\partial_{xx}\rho -2\psi\,\partial_x(g\rho)\,\partial_x\rho\;. \tag{4.2.37}$$

It is clear from the definition of the immigration and emmigration processes that $\phi>0$. Also, since we are discussing attractive random walkers $\psi>0$.[6] We see then that the effect of including these additional processes in our random walk has been to reduce the nonlinear nature of the diffusion coefficient. If ϕ and 2ψ are of the same order, i.e., $\phi=2\psi$, then we may reduce (4.2.37) to

$$\partial_t\rho \;=\; D\,\partial_{xx}\rho -2\psi(\partial_x\rho)\partial_x(g\rho) \tag{4.2.38}$$

which is the equation introduced in Montroll and West (1973). We have asumed in obtaining (4.2.38) that the diffusion from a point due to the combined effect of immigration and emmigration just cancels the diffusion to that point due to the attraction of the random walkers in the system to that location.

It is clear that if we define a function of $F(\rho)$ as,

$$F(\rho) \;=\; -2\psi\,\partial_\rho\,(\rho g) \tag{4.2.39}$$

then (4.2.38) becomes

$$\partial_t\rho \;=\; D\partial_{xx}\rho +F(\rho)(\partial_x\rho)^2 \tag{4.2.40}$$

Equation (4.2.40) is a member of a class of nonlinear differential equations which can be solved by means of a nonlinear transformation [see e.g. Montroll and West (1979)]. The solutions given by special choices of the weighting function can be interpreted in the context of our random walk model.

The solution to (4.2.40) is obtained by considering the nonlinear transformation of the dependent variable ρ,

$$u = q(\rho) \tag{4.2.41}$$

where the function q is presumably specified. We assume that $u(x,t)$ is a solution of the simple diffusion equation so that using (4.2.41) we obtain

$$\partial_t \rho = D \, \partial_{xx} \rho + D [\partial_\rho \ln[\partial_\rho q(\rho)]](\partial_x \rho)^2 \tag{4.2.42}$$

comparing (4.2.42) with (4.2.40) we have

$$F(\rho) = \partial_{\rho\rho} q(\rho)/\partial_\rho \, q(\rho) \tag{4.2.43}$$

then by inverting the process we obtain

$$u(x,t) = q(\rho) = \int^\rho \, d\rho' \, \exp \left\{ D^{-1} \int_{\rho'} F(\rho'') d\rho'' \right\} . \tag{4.2.44}$$

With $q(\rho)$ known, one can invert the function to find,

$$\rho = q^{-1}(u) \quad , \tag{4.2.45}$$

and using (4.2.44) with a given $F(\rho)$ one obtains the solution to (4.2.40). We now consider two interesting cases:

(a). The "Big Brother" Weighting Function

A form of the interspecies interaction which has an interesting interpretation is

$$g(\rho) = -\frac{\alpha}{\rho} \log(\rho/\beta) \quad ; \quad \alpha,\beta > 0 \tag{4.2.45}$$

which we term the "Big Brother" interaction. This names comes to mind because for low concentrations there is a strong bias to escape from one associates. When $\rho = \beta$ is a constant, the bias disappears and for $\rho > \beta$ there is a tendency to cluster. This set of policies would be associated with the idea that the lower the concentration of one's fellows, the more one is being watched and oppressed, but, as the population density builds up, the anonymity of the crowd makes life more bearable. This function actually has the characteristic of changing a repulsive random walk into an attractive random walk after the concentration exceeds the threshold value $\rho = \beta$.

Our differential equation has the form, using (4.2.38) and $\psi = -\eta < 0$

$$\partial_t \rho = D \, \partial_{xx} \rho - \frac{2\eta\alpha}{\rho} \, (\partial_x \rho)^2 \qquad (4.2.46)$$

so that the coefficient of the quadratic nonlinearity is

$$F(\rho) = -\frac{2\eta\alpha}{\rho} \qquad (4.2.47)$$

which can be substituted into the general solution (4.2.45) and (4.2.44). If we make the choice of parameters $\eta\alpha = D$ then we obtain, using Equation (4.2.44)

$$\rho(x,t) = \left[\int_{-\infty}^{\infty} \frac{e^{-(x'-x)^2/4Dt}}{2\sqrt{\pi Dt}} \, \frac{dx'}{\rho_0(x')} \right]^{-1} \qquad (4.2.48)$$

where $\rho_o(x') = \rho(x',0)$.

If the initial distribution of walkers is a constant, then Equation (4.2.48) yields the same constant for all time. If, however, we choose a form of the initial distribution which has the characteristic shape of a growth curve, e.g.,

$$\rho_o(x) = \frac{\rho_0}{1 + e^{-\gamma x}} \, , \qquad (4.2.49)$$

then upon integrating (4.2.48) we obtain

$$\rho(x,t) = \frac{\rho_0}{1 + e^{-\gamma(x - \gamma Dt)}} \, . \qquad (4.2.50)$$

Equation (4.2.50) describes a population front which is propagating at a velocity

$$\frac{dx}{dt} = \gamma D \, . \qquad (4.2.51)$$

(b) An "Abhorrence of Congestion" Weighting Function

A class of weighting functions which has this "abhorrence of congestion" character are those which are constant for low levels of the concentration and which increase until $g(1) \rightarrow \infty$ as the concentration reaches a saturation level unity. An example of such a function is

$$g(\rho) = -\frac{\alpha}{\rho} \, \log \, (1 - \rho) \qquad (4.2.52)$$

which has the corresponding nonlinear diffusion equation,

$$\partial_t \rho = D \, \partial_{xx} \rho + \frac{2\eta\alpha}{1 - \rho} \, (\partial_x \rho)^2 \qquad (4.2.53)$$

to solve (4.2.53) we again use (4.2.44), but with

$$F(\rho) = \frac{2\eta\alpha}{1 - \rho} \, . \qquad (4.2.54)$$

If we make the choice of parameters $2\eta\alpha = D$, then our integral expression is,

$$u(x,t) = -\frac{1}{2\sqrt{\pi Dt}} \int_{-\infty}^{\infty} e^{-(x'-x)^2/4Dt} \ln[1-\rho_0(x')] \, dx' \qquad (4.2.55)$$

where again $\rho_0(x') = \rho(x',0)$ and $u(x,t)$ is a solution to the diffusion equation in an infinite domain.

The initial distribution of concentration is selected such that the qualitative behavior of the preceding example is maintained. A distribution which does this in the present case is

$$\rho_0(x') = [1-\exp(-\epsilon e^{-\gamma x'})] \qquad (4.2.56)$$

where with $\beta > 0$ and $1 > \epsilon > 0$,

$$\rho_0(x') = \begin{cases} 1 & \text{as} \quad x' \to -\infty \\ (1-e^{-\epsilon}) & \text{as} \quad x' \to +\infty \end{cases} \qquad (4.2.57)$$

Substituting (4.2.56) into (4.2.55) and integrating yields for the concentration,

$$\rho(x,t) = \left\{ 1 - \exp[-\epsilon \, e^{-\gamma(x-\gamma Dt)}] \right\} \qquad (4.2.58)$$

which we see is again a propagating diffusion front for the population with a velocity given by (4.2.51).

4.3   Stochastic Differential Equations

There is a widely held belief that the influence of fluctuations, the rapid random variation of the environment, on the system of interest is on the whole rather trivial. One often hears the argument that the equations describing the development of a system adjust quickly to the variations in the background and are therefore really average equations capturing the *necessary* evolution of the system with the fluctuations smoothed over. The argument generally continues by making the point that the fluctuations do introduce an uncertainty in the evolution, but only in the sense of smearing the phase space orbit over some narrow region centered on the average orbit. This average orbit coincides with the deterministic evolution of the system. For a *linear* system this picture is exact, as was shown by Onsager and Machlup (1953), and portrays fluctuations as a disorganizing influence, but of secondary importance in the evolution of a system. This well-established intuitive picture turns out to be *wrong* in general.

As pointed out by Horsthemke and Lefever (1984) in their excellent monograph on the effects of fluctuations in nonlinear systems, systematic theoretical and experimental investigations have shown something quite different. In point of fact it has been found that increased environmental variability can lead to the formation of structures in nonlinear systems; structures that has no deterministic analog. These are the *dissipative structures* mentioned earlier; see e.g. Glansdorff and Prigogine (1971), and/or Nicolis and Prigogine (1977), where noise (fluctuations) induce a fundamental change in the observed structure of the system which, by analogy with the three phases of matter, is called a phase change. In this subsection we examine some of the properties of the new states to which *noise-induced transitions*[5] give rise.

4.3.1 Multiplication Fluctuations

Let us now consider a modeling strategy that is at least superficially more closely akin to the traditional ideas of fluctuations and see how environmental randomness is coupled to the evolution of systems. Recall that Verhulst modified Malthus' equation by including a quadratic nonlinearity. In general the equation that Verhulst wrote down was

$$\dot{N}(t) = k(t)\,N(t) - \Theta N^2(t) \tag{4.3.1}$$

where we now choose to make the rate parameter time dependent. Assuming that $k(t)$ can be written as the sum of a constant (average) part k_0 and a time varying (fluctuating) part $\kappa(t)$ we rewrite (4.3.1) as

$$\dot{N}(t) = k_0 N(t)\,[1 - N(t)/M] + \kappa(t)N(t) \tag{4.3.2}$$

where we have defined $\Theta M = k_0$ so that in the absence of fluctuations ($\kappa = 0$) (4.3.2) reduces to the logistic equation (3.3.3). Equation (4.3.2) models the influence of factors such as the availability of food, shelter, etc. as time dependent random variations in the rate of population growth. We can study the change in the logistic growth of the population due to the random influence of the environment by making various assumptions about the statistical properties of $\kappa(t)$. In Figure (4.3.1) we show the population of sheep in Tasmania and point out its variation around the logistic growth curve, presumably due to such environmental changes.

The first element worthy of note is that (4.3.2) is a *nonlinear stochastic differential equation*. It is unlike the traditional form of the Langevin equation studied in §2.3 in that the fluctuations enter as the coefficient of a state variable; here the state variable is the population level $N(t)$. If the fluctuation is written in the form of a linear Langevin equation then $f(t) = \kappa(t)N(t)$, i.e., the fluctuating "force" is dependent on the state of the system. When the mean level of the population is low

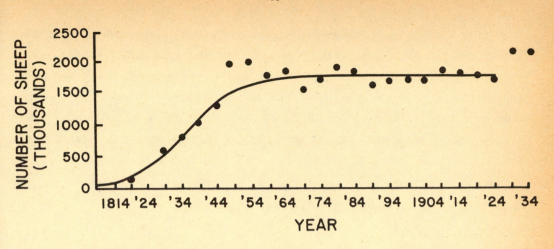

Figure 4.3.1.   Variation of sheep population in Tasmania [from J. Davison (1938)].

$(<N(t)>\ll M)$ large excursions in the rate constant are suppressed. On the other hand, when the mean level of the population becomes significant ($<N(t)>\sim M$) then even small fluctuations in $\kappa(t)$ are amplified and may modify the further evolution of the population substantially. This is a quite different effect from the state-independent fluctuations entering the linear Langevin equation studied in §2.3, where the fluctuations have essentially the same effect on the population at all stages of evolution.

We can solve (4.3.2) by dividing the equation by $N(t)$ and introducing the new variable $X(t) = \ln[N(t)/M]$ so that (4.3.2) can be rewritten as

$$\dot{X}(t) = G(X) + \kappa(t) \tag{4.3.3}$$

where G is the growth function

$$G(X) = k_0[1 - e^{X(t)}] \quad . \tag{4.3.4}$$

The advantage of (4.3.3) over (4.3.2) is that the fluctuations in the latter equation are independent of the state variable $X(t)$. Thus, even though G is a nonlinear function of X, we may be able to use the Fokker-Planck equation to determine the properties of the system. This can be done when the memory of the fluctuations is very short compared with the life expectancy of an individual or with the period between generations. Assuming the fluctuations to be Gauss distributed, zero-centered

$$<\kappa(t)> = 0 \tag{4.3.5a}$$

and delta-correlated in time

$$<\kappa(t_1)\kappa(t_2)> = 2D\delta(t_1 - t_2) \quad , \tag{4.3.5b}$$

the equation of evolution for the probability density $P(x;t\,|\,x_0)$ is the Fokker-Planck equation:

$$\partial_t P = -\partial_x \left\{ G(x) - D\,\partial_x \right\} P \tag{4.3.6}$$

which is solved subject to the initial condition $\lim\limits_{t\to 0} P(x;t\,|\,x_0) = \delta(x - x_0)$. This equation for the Verhulst case was first derived by Leigh (1969).

Although the general time dependent solution to (4.3.6) is unavailable, the steady state distribution P_{ss} can be obtained by setting $\partial_t P = 0$ to obtain

$$P_{ss}(x) = \lim\limits_{t\to\infty} P(x;t\,|\,x_0) \quad . \tag{4.3.7}$$

Note that the steady state distribution is independent of the initial state of the system and must satisfy the equation

$$\partial_x \left\{ -G(x)P_{ss} + D\,\partial_x\,P_{ss} \right\} = 0 \quad . \tag{4.3.8}$$

Therefore, when the flux of the probability through the boundaries of the interval is zero, the solution to (4.3.8) is given by

$$P_{ss}(x) = C \exp\left\{ \frac{1}{D} \int^{x} G(x')dx' \right\} \tag{4.3.9}$$

where C is the normalization constant.

We will not pursue the solution to this equation any further, since its detailed form is not of particular interest here. What does concern us here is the tactic of transforming to a new independent variable, i.e., $N(t) \rightarrow X(t)$, in order to make the stochastic differential equation additive in the fluctuations. Since it is not always possible to do this, especially in systems with more than one degree of freedom, we examine the properties of more general systems in which fluctuations are state-dependent. Consider the stochastic differential equation

$$\dot{N}(t) = G(N) + g(N)\,\kappa(t) \tag{4.3.10}$$

where $G(N)$ and $g(N)$ are both analytic functions of the population N and $\kappa(t)$ is a stochastic parameter whose statistical properties need to be specified. The state-dependent fluctuations have been assumed to be factorable, that is to say that $g(N)$ is a deterministic function, and when written in this form the fluctuations are referred to as multiplicative. The specific forms of $G(N)$ and $g(N)$ are usually obtained by phenomenological reasoning for the particular process being investigated although some systematic techniques have recently been developed, see e.g., van Kampen (1981) and Lindenberg and West (1984). In the population growth problem above we choose $G(N)$ to be the logistic growth function $k_0 N(1 - N/M)$ and $g(N)$ to be the population level N.

Equation (4.3.10) encapsulates a number of phenomenological equations depending on how the functions G and g are interpreted. In dynamics G is the derivative of a potential, i.e., a force law; in hydrodynamics it is a convective derivative plus dissipation; in chemical kinetics it is the law of mass action; in population biology it is the saturation-inducing growth law; in short, G *gives the deterministic evolution of the system in isolation from the environment.* The equation $\dot{N} = G(N)$ is the theoretical macroscopic equation of motion. There is often an underlying theory to provide guidance in the construction of $G(N)$. There is all of analytic dynamics at our disposal if we want to put a space station in orbit or describe the behavior of a gas; physical chemistry tells how to combine elementary particle composites to describe complex reactions; but there is little guidance for constructing the function $g(N)$. This function describes the coupling between the known and the unknown, that is, between the deterministic evolution of a system and the modification of that evolution due to the coupling of the system to the environment.

Let us again consider the modified Verhulst equation (4.3.2), only this time we transform the variable to $X(t) = N(t)/M$. The Fokker-Planck equation directly corresponding to (4.3.2) is[6]

$$\partial_t P = -k_0 \, \partial_x \left[x(1-x)P \right] + D \, \partial_x \left[x \, \partial_x \, (xP) \right] \tag{4.3.11}$$

again we cannot obtain the full time dependent solution to this equation, but we can construct the steady state equation

$$\partial_x \left\{ -k_0 x(1-x)P_{ss} + Dx \, \partial_x \, (x\,P_{ss}) \right\} = 0 \tag{4.3.12}$$

which has the normalized solution

$$P_{ss}(x) = C\,x^{\frac{k_0}{D}-1}\,e^{-x/D} \quad . \tag{4.3.13}$$

The steady-state solution is normalizable for $k_0 > D$, and therefore is a proper probability density with

$$C = D^{k_0 D}/\Gamma(k_0/D) \quad . \tag{4.3.14}$$

We can also rewrite (4.3.13) in the form

$$P_{ss}(x) = C\exp\left\{ -\frac{x}{D} + \left(\frac{k_0}{D} - 1\right)\ln x \right\} \tag{4.3.15}$$

and note the following facts: 1) the modes (maxima) of a probability density function indicate the *most probable* states of the system; 2) the relative heights of the maxima indicate the relative probability of their occurrance and 3) they may be associated with the macroscopic steady-states of the system. The maxima of (4.3.15) are found as in §2.1 by taking the derivative of the term in the exponential and setting it to zero:

$$\partial_x \left\{ -\frac{x}{D} + \left(\frac{k_0}{D} - 1\right)\ln x \right\} = 0 \quad . \tag{4.3.16}$$

Thus the maxima are the roots of equation (4.3.16):

$$x_m = k_0 - D \quad . \tag{4.3.17}$$

If we examine the deterministic equation (4.3.2), however, we see that the equilibrium equation $\dot{x} = 0$ has the solutions

$$x_{ss} = 0 \ \text{ or } \ 1 \quad . \tag{4.3.18}$$

Thus the maximum of the probability distribution given by (4.3.17) is a qualitatively new state of the system, distinct from the steady state solutions of the deterministic system. This qualitative change has been induced by the multiplicative fluctuations and denotes the *fluctuation induced nonequilibrium phase transition.*

The extrema of the steady state distribution have a particularly interesting interpretation in terms of the measurements one makes on a system. If the transients in the system have died away so that the steady state distribution is a faithful description, then $P_{ss}(x)dx$ is the fraction of the time the system is in the interval $(x, x + dx)$ over a long period of time. This is another way of saying the system is ergodic, i.e., the long time average and steady state ensemble average are equivalent. Therefore the maxima of $P_{ss}(x)$ indicate the state in which the system spends most of its time relative to the full range of possible values of x. These are the states most likely to be observed in a laboratory experiment or a field study. In an analogous way the system is "repelled" by the minima of the steady state distribution. In the example of the Verhulst equation there was a single maxima, but in general the steady state distribution has the form

$$P_{ss}(x) = C\exp\{-U(x)/D\} \tag{4.3.19}$$

where C is the normalization constant and $U(x)$ is the *stochastic potential:*

$$U(x) = -\int^{x} \frac{G(x')}{g^2(x')} \ dx' + D\ln g(x) \quad , \tag{4.3.20}$$

here (4.3.10) is the stochastic differential equation for which (4.3.19) is the steady state probability density.

When the external fluctuations are additive, i.e., the state-independent $g(x) =$ constant, and $G(x)$ can be written as the gradient of a potential $V(x)$, then the stochastic potential and the deterministic potential coincide. In this situation the extrema of the $P_{ss}(x)$ correspond to the extrema of the determinisitc potential, i.e., the points where $\frac{\partial V(x)}{\partial x} = 0$. The minima of the deterministic potential coincide with the most probable states and the maxima with the least probable states. The potential minima are the stable steady states and the potential maxima the unstable steady states, under the condition that the probability flux vanishes in the steady state. This interpretation carries over to the case of state-dependent fluctuations, but the state dependence of $g(x)$ can shift the location, type and number of extrema, so that the phases of the stochastic potential can be quite different from those of the deterministic potential. Horsthemke and Lefever (1984) summarize these points as follows:

i.   A transition occurs when the functional form of the random variable describing the steady state of the system changes qualitatively.

ii.   This qualitative change is most directly reflected by the extrema of the stationary probability law, ...

iii.   The physical significance of the extrema, apart from being the most appropriate indicator of a transition, is their correspondence to the macroscopic phases of the system. The extrema are the order parameter of the transition.

The extrema of $P_{ss}(x)$ can be found directly from the vanishing of the gradient of the stochastic potential, $\dfrac{\partial U\ (x)}{\partial x} = 0$, to yield to the relation:

$$-G(x_m) + D\ g(x_m)\ \frac{\partial g(x)}{\partial x}\ \big|_{x=x_m} = 0 \ .$$

(4.3.21)

This is the starting equation for discussing the number and type of extrema x_m of the nonlinear system in the steady state under the influence of rapid external fluctuations. The detailed properties of the steady state are determined by the mean square level, D, of the fluctuations. If D is sufficiently small then the properties of the system are not much different from the deterministic case. When the intensity of the fluctuations becomes appreciable, say D exceeds some threshold level, then a transition occurs. In this latter case, in addition to the disorganizing effect that *any* fluctuation has on a system, the multiplicative fluctuations create new states which are not expected on the basis of the usual phenomenological descriptions.

[1]The concept of wave-particle duality has its origin in quantum mechanics and reflects the idea that an electron (or other-on) has both particle-like and wave-like properties and therefore cannot be viewed as one or the other, but is in fact a synthesis of the two.

[2]This system of equations is obtained by truncating the field equations for a two-dimensional flow field in velocity and temperature on a square grid with periodic boundary conditions. If the fluid has a uniform depth H, the temperature difference between the upper and lower surfaces is maintained at a constant value ΔT, g is the acceleration of gravity, α is the coefficient of thermal expansion, ν is the kinematic viscosity and κ is the coefficient of thermal expansion, then the *Rayleigh number* is $Ra = g\alpha H^3 \Delta T/\nu\kappa$. The critical Rayleigh number is $R_c = \pi^4(1+a^2)^3/a^2$. Where a is the length of the box, the dimensionless time in (4.1.1) is $\tau = \pi^2(a^2+1)\kappa t/H^2$, $\sigma = \nu/\kappa$ is the *Prandtl number*, $r = R_a/R_c$ and $b = 4/(1+a^2)$. In these equations X is proportional to the intensity of the convective motion, while Y is proportional to the temperature difference between the ascending and descending currents and Z is proportional to the distortion of the vertical temperature profile from linearity.

[3]Here we use the terms random, chaotic, stochastic etc. interchangeably. Later we will be more pedantic in their use.

[4]We note that if we were considering repulsive random walkers so that $\phi<0$ then the definition of immigration and emmigration are interchanged and $\phi<0$.

[5]Horsthemke and Lefever (1984) repeatedly use the term *noise-induced nonequilibrium phase transition* to emphasize the important aspects of these process.

[6]We note that if we were considering repulsive random walkers so that $\psi<0$ then the definition of immigration and emmigration are interchanged and $\phi<0$.

5. WHAT IS IT THAT WAS REALLY SAID?

The series of lectures on which this essay is based was developed for an audience with a heterogeneous scientific background which varied from music theory to psychiatry to organic chemistry. The intent of the series was to draw parallels between different disciplines and to perhaps uncover an underlying unity. Instead of this, we in fact uncovered the self-imposed limitations that scientific investigators have established in their efforts to make sense out of complex systems and processes. Starting with the first of the modern scientists (Newton) I have attempted to show how the evolution of a linear world view has dominated scientific thinking in natural philosophy. The efforts of Newton to understand mechanistic relations and those of Gauss to interpret large data sets, along with their contemporaries, have contributed to this limited and limiting view of the world. It is not my intent to in any way minimize the monumental contributions of these two great men to our understanding of the physical world. But I do wish to emphasize the influence that this style of thought has had outside the physical sciences. For example, it was this perspective that prompted Descartes to say:

> "If we really knew all the parts of the seed of any particular animal, man, for example, we could deduce from that alone, by certain and mathematical reasoning, the shape and conformation of each of his limbs."

In the various lectures I have attempted to indicate some of the limitations of this reductionistic point of view. Firstly, in this view there *appears* to be a direct causal chain that implicitly (or explicitly) leads from the microscopic to the macroscopic. This direct chain idea relies on a linear relation between the observed macroscopic phenomenon and a deeply concealed microscopic cause. One example of the success of this viewpoint is the explanation of "Down's Syndrome," a particular form of mental retardation that has been traced to a specific genetic defect (an extra chromosome). In fact it may be quite *possible* that one can trace other abnormal or pathological states to similar aberrations in the genetic and/or micro-chemical makeup of individuals. However, there has been a dramatic lack of similar successes in the resolution of the structure-function question in biology. Thus it *might* be possible to identify a microscopic cause of a particular disruption of a normal phenomenon without understanding the *cause* of the phenomenon itself.

Secondly, the reductionistic view requires that one can in principle know (measure) the connection between the microscopic and macroscopic. However, as we have discussed, the observable consequences and patterns in a complex system are not necessarily uniquely related to any particular microscopic state, but are rather related to an ensemble of these unobserved states. Consequently the characterization of the phenomenon is through a probability distribution function rather than by means of deterministic dynamics. This uncertainly is not always a result of the course graining of the measurements preformed, although it often is, but can result from the nonlinear properties of the system itself. The failure of the Gauss distribution to capture the random variations in many data sets illustrates the importance of such nonlinearities. The dictum of Newton in the introduction to his *Optics*:

"My design in this book is not to explain the properties of Light by

Hypothesis but to propose and prove them by Reason and Experiments."
remains applicable today so that when the data so indicate the hypothesis of linearity must be abandoned. Thus the often observed form of a power-law distribution in the social and behavioral sciences strongly suggests that the underlying process is not linear.

Thirdly, the measurable effect is often the result of a strong nonlinear interaction that results in a coherent (deterministic) state that is apparently not dependent on the form of the interaction among the constituent microscopic elements. The use of limit cycles in the mathematical description of biorhythms is an example of this, in that the limit cycle itself describes the biological process. The coupling of the relaxation oscillator to the environment acts to stabilize the dynamic process so that the excursion of the oscillator is neither too large nor too small, and the periodic rhythm of the beating of the heart, for example, emerges. Thus the macroscopic degrees of freedom organize the way in which the microstructure of the system regulates the observable phenomenon, i.e., there is a feedback loop that stabilizes the biorhythm. Because of this it is not possible to deduce the microstructure from observations of the organized dynamic state while it is operating correctly because the dynamics are *not* determined at the microscopic level. A description of the organizing process may however provide some indications of the microscopic mechanisms.

Fourthly, we have seen that the reduction of the description of the system dynamics to a low order deterministic set of equations is not sufficient to ensure the predictability of its evolution. The phenomena of *deterministic chaos* implies that even "simple" nonlinear systems can have a rich variety of both deterministic and apparently random dynamics. Therefore it would be in everyone's best interest if it were fully appreciated that in natural philosophy a simple system does not necessarily possess simple dynamic properties. Also that observed complexity does not imply a corresponding complexity in the underlying dynamical rule. Thus the fluctuations in a data set may not necessarily indicate the disruptive random effect of the environment nor that of sampling error, but may instead be a consequence of an inflexible nonlinear rule for the system's evolution.

I conclude that the robustness of certain linear models has little or nothing to do with the typical processes found in natural phenomenon outside the physical sciences. The stability of the patterns that characterize social and behavioral processes are more often than not a consequence of the nonlinear nature of the process, i.e., they are manifestations of a structural stability in the system. This structural stability appears, for example, in the phase space orbits of relaxation oscillators and in the spatial patterns that arise in certain reaction-diffusion chemical systems. The importance of understanding the properties of such systems has been appreciated for quite some time, but it is only within the past decade or so that the synthesis of mathematical analysis and computational techniques for computers has enabled scientists to break the linear chain of logic that had encased their mathematical models.

We have now emerged from the forest of nonlinear mathematics with, I hope, a greater sense of the correspondence between the richness of the mathematics and the complexity of the processes in natural philosophy. As a closing observation I wish to emphasize that complex systems are intrinsically nonlinear and stochastic, perhaps one or the other dominating in a particular situation, but in general both aspects are important for a complete understanding of the system. How one utilizes this insight in a particular context remains to be seen.

APPENDIX - COMPUTER PROGRAMS

1. RANDOM WALK I
   RANDOM WALK II

2. TIME SERIES

3. LEVY WALKS

1. Random Walk I

In this program we present a walker taking random steps on a two dimensional plane. Two random numbers are generated, one for the right-left step choice and the other for the up-down step choice. The program will run until manually stopped.

Random Walk II

The program written below is for a random walker who choses to take a unit step to the right, left or remain stationary based on whether the value of a random number is in the internal $(0, 1/3)$, $(2/3, 1)$ or $(1/3, 2/3)$, respectively. The number of times the walker occupies a given interval on the x-axis is accumulated and plotted as a histogram [cf. Figure (2.1.4)].

2. Time Series

Here we present a simple program to generate the Fourier series representation of a time series. The mode amplitudes $A(P)$ are here given an explicit functional form, but they can also be read in as data; similarly for the frequency $OM(P)$. Try a run with $N1 = 200$, $OMEG = 0.05$ and $A = 1.5$; then compare runs with increasing number of modes starting with $N2 = 1$. A phase $\phi(P)$ can also be put into the cosine term and the resulting form of the time series for various choices of these initial phases examined.

3. Lévy Walks

As in RANDOM WALK I in this program we present a walker taking random steps on a two dimensional plane. The distribution of step sizes is given by a power law with index ALP (here $= 1.3$) and the rotation angle for the step is uniform in the interval $(0, 2\pi)$. The number of steps in the walk is set by N.

REFERENCES

M.J. Ablowitz and H. Segur, *Solitons and the Inverse Scattering Transform,* SIAM, Philadelphia (1981).

V.I. Arnold, *Ordinary Differential Equations,* MIT Press, Cambridge, Mass. (1981).

L. Bachelier, Annales Scientifiques de l'Ecole Normale Supérieure, Sup 3, No. 1017 (1900).

W.W. Badger, in *Mathematical Models as a Tool for the Social Sciences*, ed. B.J. West, Gordon and Breach (1980).

E. Basar, *Biophysical and Physiological Systems Analysis*, Addison-Wesley Reading, Mass. (1976).

Y. Beers, *Introduction to the Theory of Errors*, Addison-Wesley, Reading, Mass. (1953).

J. Bellet, *Clinical Disorders of the Heartbeat*, 3rd ed., Lea and Febiger, Philadelphia (1971).

D. Bernoulli, *Reflexions et Eclaircissements sur les Nouvelles Vibrations des Cordes Exposes dans les Memoires de l'Academie*, Roy. Acad. Berlin, 147 (1755).

M.V. Berry, J. Phys. A: Math. Gen. **12**, 781 (1979).

M.V. Berry and Z.V. Lewis, Proc. Roy. Soc. Lond. **370A**, 459 (1980).

P.R. Bevington, *Data Reduction and Error Analysis for the Physical Sciences*, McGraw-Hill, St. Louis (1969).

K.E. Boulding, Forward to *Population: The First Essay*, by T.R. Mathus, Univ. Mich. Press, Ann Arbor (1959).

L. Brillouin, *Wave Propagation in Periodic Structures*, 2nd Ed. Dover, New York (1946).

R. Brown, Phil. Mag. **6**, 161 (1829); Edinburgh J. Sci. **1**, 314 (1829).

A. Cauchy, Comptes Rendus **37**, 202 (1853).

S. Chandrasekhar, Rev. Mod. Phys. **15**, 1 (1943).

S. Chapman, Phil. Trans. Roy. Soc. **A216**, 279 (1916).

D.L. Cohen, Bull. Math. Biophysics **16**, 59 (1954); ibid **17**, 219 (1955).

J.S. Coleman, *Introduction to Mathematical Sociology*, The Free Press of Glencoe, Collier-Macmillan, London (1964).

P. Collet and J.P. Eckmann, *Iterated Maps on the Interval as Dynamical Systems*, Birkhäuser, Basel (1980).

P. Cootner, editor, *The Random Character of the Stock Market*, MIT Press, Cambridge, MA (1964).

E. Cortes, B.J. West and K. Lindenberg, J. Chem. Phys. **82**, 2708 (1985).

J.H. Curry, Comm. Math. Phys. **68**, 129 (1979).

J. d'Alembert, Recherches sur la Courbe que Forme une Corde Tendue Mise en Vibration, Roy. Acad. Berlin, (1747).

C. Darwin, *The Origin of the Species by Means of Natural Selection on the Preservation of Favored Races in the Struggle of Life* (1859).

C. Darwin, Letter to J.M. Herbert, 1844 or 1845, in *The Life and Letters of Charles Darwin*, F. Darwin, ed., vol. **1**, p. 334, London (1887), Murray.

J. Davidson, Trans. Roy. Soc. South Aus. **62**, 342 (1938).

A. de Moivre, *Approximatio ad Summam Taerminorum Binomii* $\overline{(a+b)^n}$, (1733).

A. de Moivre, *Doctrine of Chances; or A Method of Calculating the Probabilities of Events in Play* (London, 1718). 3rd ed. (1756) printed by Chelsea Press (1967).

Descartes, Formation de l'animal, Oeuvres, vol. XI, 277.

D. de Sola Price, *Little Science, Big Science,* Columbia Univ. Press, N.Y. (1963).

Dirichlet, J. für Math. IV, 157 (1829).

L. Dossey, *Space, Time and Medicine*, Shanbhala, Boulder and London (1982).

J.P. Eckmann, Rev. Mod. Phys. **53**, 643 (1981).

A. Einstein, Ann. Physik **17**, 549 (1905).

D.K. Faddeev, *Mathematics* vol.3, ed. A.D. Aleksandrov, A.N. Kolmogorov and M.A. Lavrentev, MIT Press, Cambridge (1964).

D. Farmer, J. Crutchfield, H. Froehling, N. Packard and R. Shaw, in *Nonlinear Dynamics*, ed. R.H.G. Helleman, Ann. N.Y. Acad Sci. **357**, 453 (1980).

M.J. Feigenbaum, J. Stat. Phys. **19**, 25 (1978); J. Stat. Phys. **21**, 669 (1979).

S.D. Feit, Comm. Math. Phys. **61**, 249 (1978).

W. Feller, Acta Biotheoretica **5**, 51 (1940).

E. Fermi, J. Pasta and S. Ulam, in *Collected Works of Enrico Fermi*, vol. II, 978, Chicago (1965).

J.C. Fisher and R.C. Pry, in *Practical Applications of Technological Forecasting in Industry*, ed. M.J. Cetron, Chichester, New York (1971).

R.A. Fisher, Proc. Roy Soc. Edin. xlii, 321 (1922).

R.A. Fisher, *The Genetic Theory of Natural Selection*, Dover, New York (1929); 2nd rev. ed. (1958).

R.A. Fisher, Ann. Eugen. **7**, 355 (1937).

G.W. Ford, M. Kac and P. Mazur, J. Math. Phys. **6**, 504 (1965).

A.D. Fokker, Ann. Physik **43**, 1812 (1914).

J. Fourier, Théorie Analytique de la Chaleur, Paris (1822).

S. Freud and Breuer, *Studies in Hysteria*. (1895).

U. Frisch, P.L. Sulem and M. Nelkin, J. Fluid Mech. **87**, 719 (1978).

F. Gauss, *Theoria motus corporum coelestrium,* Hamburg (1809); Dover Eng. Trans.

J. Gillis and G.H. Weiss, J. Math. Phys. **11**, 1308 (1970).

P. Glansdorff and I. Prigogine, *Thermodynamic Theory of Structure, Stability and Fluctuation,* Wiley, New York (1971).

L. Glass, M.R. Guevara and A. Shrier, Physica **7D,** 89 (1983).

B.V. Gnedenko and A.N. Kolmogorov, *Limit Distributions for Sums of Independent Random Variables,* Addison-Wesley, Reading, Mass. (1954).

N. Goel, S. Maitra and E.W. Montroll, Rev. Mod. Phys. **43**, 231 (1971).

A.L. Goldberger and E. Goldberger, *Clinical Electrocardiography*, The C.V. Mosbey Comp, St. Louis (1981).

R. Gompertz, Philos. Trans. Roy. Soc. London **115**, 51 (1825).

M.R. Guevara and L. Glass, J. Math. Bio. **14**, 1 (1982).

G.H.L. Hagen, *Grundzüge der Wahrscheinlichkeitsrechnung,* Berlin, (1837); 2nd Edition (1867).

H. Haken, *Synergetics, An Introduction*, Springer-Verlag, Berlin (1978)

G.H. Hardy, Quart. J. Math. **38**, 269 (1907).

G.H. Hardy, Trans. Am. Math. Soc. **17**, 301 (1916).

W. Harvey, *On the Motion of the Heart and Blood in Animals* (1628), H. Rognery Co., Chicago (1962).

M. Henon, Comm. Math. Phys. **50**, 69 (1976).

W. Horsthemke and R. Lefever, *Noise-Induced Transitions*, Springer-Verlag, Berlin (1984).

B.D. Hughes, M. Shlesinger and E.W. Montroll, Proc. Nat. Acad. Sci. USA **78**, 3287 (1981).

B.D. Hughes, E.W. Montroll and M.F. Shlesinger, J. Stat. Phys. **28**, 111 (1982)

B.D. Hughes, E.W. Montroll and M.F. Shlesinger, J. Stat. Phys. **30**, 273 (1983).

J. Ingen-Housz, *Dictionary of Scientific Biology*, ed. C.C. Gillispie, Scribners, New York, p. 11 (1973).

M. Jammer, *Concepts of Force*, Harper and Row (1960).

G. Jona-Lasinio, Nuovo Cimento **26B**, 99 (1975).

S. Jorna, editor, *Topics in Nonlinear Dynamics,* AIP Conf. Proc. **46**, (1978).

J. Kemeny and J.L. Snell, *Mathematical Models in the Social Sciences*, MIT Press, Cambridge, Mass. (1972).

A. Ya Khinchine and P. Lévy, Comptes Rendus **202**, 274 (1936).

A.N. Kolmogorov, Math. Ann. **104**, 415 (1931)

A.N. Kolmogorov, Giorale Ist. Ital. Attuari **7**, 74 (1936).

A.N. Kolmogorov, I. Petrovsky and N. Piscounoff, Bull. de l'Univ. d'Etat à Moscou (Sér. Internat.) A., **I**, 1 (1937).

E.J. Kormandy, *Concepts of Ecology*, pp. 95-97 Prentice-Hall, New Jersey (1969).

J.L. Lagrange, Recherches sur la Nature et la Propagation du Son, Miscellanea Taurinesiu, t. I (1759).

O.E. Lanford, in *Statistical Mechanics and Dynamical Systems*, Mathematics Dept., Duke Univ., Durham, N. Car., Chap. 4 (1976).

P. Langevin, Comptes Rendus Acad. Sci. Paris, 530 (1908).

P.S. Laplace, *Traité de Mécanique Celeste*, Paris (1825).

M.A. Lavrentev and S.M. Nikol'skii, in *Mathematics vol. 1*, eds. A.D. Aleksandrov, A.N. Kolmogorov and M.A. Lavrentev, MIT press, Cambridge (1964).

E.G. Leigh, *Some Math Problems in Biology 1*, Am. Math. Soc. (1969).

P. Lévy, *Calcul des probabilités*, Gauthier-Villars, Paris (1925).

P. Lévy, *Théorie de l'addition des variables aléatoires*, Gauthier-Villars, Paris (1937).

T-Y Li and J.A. Yorke, Ann. Math. Monthly **82**, 985 (1975).

K. Lindenberg, K. Shuler, V. Seshadri and B.J. West in *Probabilistic Analysis and Related Topics*, Vol. 3, ed. A. T. Bharucha-Reid, Academic Press, New York (1983).

K. Lindenberg and B.J. West, J. Atmos. Sci., **41**, 3021 (1984).

R.B Lindsay, in *The Theory of Sound,* vol.1, J.W.S. Rayleigh, Dover (1945).

A.J. Lichtenberg and M.A. Lieberman, *Regular and Stochastic Motion*, Springer-Verlag, New York (1983).

E.N. Lorenz, J. Atmos. Sci. **20**, 130 (1963).

A.J. Lotka, J. Wash. Acad. Sci. **16**, 317 (1926).

A.J. Lotka, *Elements of Mathematical Biology,* Williams and Wilkins (1925); Dover (1956).

G.G. Luce, *Biological Rhythms in Human and Animal Physiology*, Dover, New York (1971).

D.A. MacLuich, *Biological Studies*, **43** (1937).

C.D. Majumdar, Phys. Rev. **82**, 844 (1951).

R.T. Malthus, *Population: The First Essay* (1798), Univ. Mich. Press, Ann Arbor (1959).

B.B. Mandelbrot, CR **280A**, 1551 (1975).

B.B. Mandelbrot, in *Nonlinear Dynamics*, ed. R.H.G. Helleman, Ann. New York Acad. Sci. **357**, 249 (1980).

B.B. Mandelbrot, *Fractals, Form and Chance*, W.F. Freeman, San Francisco (1977).

B.B. Mandelbrot, *The Fractal Geometry of Nature*, W.F. Freeman, San Francisco (1982).

A.A. Maradudin, E.W. Montroll, G.H. Weiss and I.P. Ipatova, *Theory of Lattice Dynamics in the Harmonic Approximation*, 2nd Ed., Academic Press, New York (1971).

R.M. May, Science **177**, 900 (1972).

R.M. May, Ann. Nat. **107**, 46 (1972).

R.M. May, Nature **261**, 459 (1976).

E.W. Montroll, Lect. Th. Phys. (Th. Phys. Inst., Univ. Col.) X-A, 531 (1968).

E.W. Montroll, Proc. Nat. Acad. Sci. USA **75**, 4633 (1978).

E.W. Montroll and W.W. Badger, *Introduction to Quantitative Aspects of Social Phenomena*, E.W. Montroll, North-Holland, Amsterdam (1984).

E.W. Montroll and B.J. West in *Synergetics*, ed. H. Haken, B.G. Teubner, Stuttgart (1973).

E.W. Montroll and B.J. West, in *Fluctuation Phenomena*, eds. J.L. Lebowitz and E.W. Montroll, North-Holland, Amsterdam (1979).

A.M. Mood, *Introduction to the Theory of Statistics*, McGraw-Hill, New York (1950).

O.E. Neugebaurer, *The Exact Sciences of Antiquity*, Princeton Univ. Press, Princeton, N.J. (1952).

I. Newton, *Principia*, Book II (1686).

G. Nicolis and I. Prigogine, *Self-Organization in Nonequilibrium Systems. From Dissipative Structure to Order Through Fluctuations*, Wiley, New York (1977).

L. Onsager and S. Machlup, Phys. Rev. **91**, 1505 (1953), ibid 1512 (1953).

S. Orey, Z. Wahrsch' theorie verw. Geb. **15**, 249 (1970).

E. Ott, Rev. Mod. Phys. **53**, 643 (1981).

N.H. Packard, J.P. Crutchfield, J.D. Farmer and R.S. Shaw, Phys. Rev. Lett. **45**, 712 (1980).

V. Pareto, *Cours d'Economie Politique*, Lausanne (1897).

K. Pearson, Nature **72**, 294 (1905).

K. Pearson, Biometrika **13**, 25 (1920).

P.J.E. Peebles, *The Large-Scale Structure of the Universe*, Princeton Univ. Press, Princeton, New Jersey (1980).

J. Perrin, *Brownian Movement and Molecular Reality*, Taylor and Francis, London (1910).

M. Planck, Sitzber. Preuss. Akad. Wiss. Physik. Math **K1**, 324 (1917).

H. Poincare, *The foundations of Science*, The Science Press, New York, (1913).

I. Procaccia and J. Ross, Science **198**, 716 (1977).

W.P. Provine, in *Mathematical Models in Biological Discovery,* eds. D.L. Solomon and C. Walter, Springer-Verlag, Berlin (1977).

N. Rashevsky, *Mathematical Biophysics Physico-Mathematical Foundations of Biology*, Vol. 2, 3rd rev. ed., Dover, New York (1960).

N. Rashevsky, in *Physicomathematical Aspects of Biology*, Int. Sch. Phys. XVI, ed. N. Rashevsky, Academic Press, New York (1962).

Lord Rayleigh, Nature **72**, 318 (1905).

E.C.R. Reeve and J.S. Huxley, in *Essays on Growth and Form*, eds. W.E. LeGros Clark and P.B. Medarvar, Oxford, Clarendon Press (1945).

L.M. Ricciardi, *Diffusion Processes and Related Topics in Biology* Lect. Notes in Biomath. **14**, Springer-Verlag, Berlin (1977).

F.S. Roberts, *Measurement Theory*, Ency. of Math. **7**, Addison-Wesley, Reading, Mass. (1979).

M. Romanowski, *Random Errors in Observation and the Influence of Modulation on Their Distribution,* Verlag Komad Wittwer, Stuttgart (1979).

O.E. Rössler, Phys. Lett. **57A**, 397 (1976).

D. Ruelle and F. Takens, Comm. Math. Phys. **20**, 167 (1971).

D. Sahal, *Patterns of Technological Innovation*, Addison-Wesley, London (1981).

J. Salk, *The Survival of the Wisest*, Columbia University Press, New York (1983).

E. Schrödinger; *What is Life? The Physical Aspects of the Living Cell*, Cambridge Univ. Press, London (1943).

A. Scott, in *Nonlinear Phenomena in Physics and Biology*, edited by R.H. Enns, B.L. Jones, R.M. Miura and S.S. Rangnekar, NATO Scientific Affairs Div. Plenum Press, N.Y. (1981).

H.L. Seal, Biometrika **54**, 1 (1967).

L.A. Segel, *Modeling Dynamic Phenomena in Molecular and Cellular Biology*, Cambridge University Press, London (1984).

V. Seshadri and B.J. West, Proc. Nat. Acad. Sci. USA **79**, 4501 (1982).

R. Shaw, Z. Naturforsch, **36a**, 80 (1981).

M.F. Shlesinger and B.D. Hughes, Physica **109A**, 597 (1981).

W. Shockley, Proc. of IRE **45**, 279 (1957).

Simó, J. Stat. Phys. **21**, 465 (1979).

R.H. Simoyi, A. Wolf and H.L. Swinney, Phys. Rev. Lett. **49**, 245 (1982).

T. Simpson, Phil. Trans. Roy. Soc. Lond. **49**, Part I, 82 (1755).

M. Smoluchowski, Phys. Zeitschrift **17**, 557 and 585 (1916); Ann. Physik **21**, 756 (1906).

D.L. Soloman and C. Walter, editors, *Mathematical Models in Biological Discovery*, Springer-Verlag, Berlin (1977).

R.L. Stratonovich, *Topics in The Theory of Random Noise, Vol. 1, 2*, Gordon and Breach, New York (1967).

Th. Svedberg, *The Existence of the Molecule*, Leipzig (1912).

Yu. M. Svirezhev and D.O. Logofet, *Stability of Biological Communities*, Mir Publ., Moscow (1983).

D.W. Thompson, in *The World of Mathematics*, vol. 2, 1001, ed. J.R. Newman, Simon and Shuster, New York, (1956).

D.W. Thompson, *On Growth and Form*, (1915), abridged ed., Cambridge (1961).

T.C. Tippett, in *The World of Mathematics*, vol. 3, 1459 ed. J.R. Newman, Simon and Shuster, New York (1956).

G.E. Uhlenbeck and L.S. Ornstein, Phys. Rev. **36**, 823 (1930).

B. van der Pol, Phil. Mag. **2**, 978 (1926).

B. van der Pol and J. van der Mark, Nature **120**, 363 (1927).

B. van der Pol and J. van der Mark, Phil. Mag. **6**, 763 (1928).

B. van der Pol and J. van der Mark, Extr. arch. neerl. physiol. de l'homme et des animaux **14**, 418 (1929).

N.G. van Kampen, *Stochastic Processes in Physics and Chemistry*, North-Holland, Amsterdam (1981).

P.F. Verhulst, Mem. Acad. Roy. Bruxelles **28**, 1 (1844).

A.A. Verveen and L.J.De Felice, in *Progress in Biophysics and Molecular Biology*, eds. A.J.V. Butter and D. Noble, Pergamon Press, Oxford (1974).

M.V. Volkenshtein, *Biophysics*, Mir Publ, Moscow (1983).

V. Volterra, Leçons sur la théorie mathématique de la lutte pour la vie, Cahiers Scientifiques VII, Paris, Gauthier-Villars, VI (1931).

M.C. Wang and G.E. Uhlenbeck, Rev. Mod. Phys. **77**, 323 (1945).

R.R. Ward, *The Living Clocks*, A.A. Knopf, New York (1971).

D. Wechsler, *The Measurement and Appraisal of Adult Intelligence*, 4th Ed. p. 107, Williams and Wilkins Co., Baltimore (1958).

E.R. Weibel, *Morphometry of the human being*, Academic Press, New York (1963).

H. Weiner, *Psychosomatic Medicine and the Mind-Body Problem in Psychiatry*, in Handbook of the History of Psychiatry, ed. E. Wallace IV, Yale University Press (1984).

G.H. Weiss and R.J. Rubin, in *Advances in Chemical Physics, Volume 52*, editors I. Prigogine and S.A. Rice, John Wiley (1983).

B.J. West, Coll. Phenom. **1**, 195 (1974).

B.J. West, editor, *Mathematical Models as a Tool for the Social Sciences*, Gordon and Breach, New York (1980).

B.J. West, A. Bulsara, K. Lindenberg, K.E. Shuler and V. Seshadri, Physica **97A**, 211 (1979).

B.J. West, A.L. Goldberger, G. Rovner and V. Bhargava, Physica D **16**, (in press).

B.J. West and K. Lindenberg, in *Predictability of Fluid Motions*, eds. G. Holloway and B.J. West, AIP Conf. Proc. **106**, 45 (1983).

N. Wiener, *Extrapolation, Interpolation and Smoothing Stationary Time Series*, MIT Press, Cambridge, Mass. (1949).

N. Wiener, *Collected Works vols. I,II and III*, MIT Press, Cambridge, Mass. (1984).

A.T. Winfree, J. Th. Biol. **16**, 15 (1977).

A.T. Winfree, Sci. Am. **249**, 144 (1984).

Yule and Kendall, *Introduction to the Theory of Statistics*, (14th Ed. 1950) Chas. Griffin and Co., London.

N.J. Zabusky and M.D. Kruskal, Phys. Rev. Lett. **15**, 241 (1965).

A. Zygmund, *Trigonometric Series, Vol. I and 2*, Cambridge University Press, Cambridge (1935).

Biomathematics

Managing Editor: S. A. Levin

Editorial Board: M. Arbib, H. J. Bremermann, J. Cowan, W. M. Hirsch,
S. Karlin, J. Keller, K. Krickeberg, R. C. Lewontin, R. M. May,
J. D. Murray, A. Perelson, L. A. Segel

Volume 14
C. J. Mode

Stochastic Processes in Demography and Their Computer Implementation

1985. 49 figures, 80 tables. XVII, 389 pages. ISBN 3-540-13622-3

Contents: Fecundability. – Human Survivorship. – Theories of Competing Risks and Multiple Decrement Life Tables. – Models of Maternity Histories and Age-Specific Birth Rates. – A Computer Software Design Implementing Models of Maternity Histories. – Age-Dependent Models of Maternity Histories Based on Data Analyses. – Population Projection Methodology Based on Stochastic Population Processes. – Author Index. – Subject Index.

Volume 13
J. Impagliazzo

Deterministic Aspects of Mathematical Demography

An Investigation of the Stable Theory of Population including an Analysis of the Population Statistics of Denmark

1985. 52 figures. XI, 186 pages. ISBN 3-540-13616-9

Contents: The Development of Mathematical Demography. – An Overview of the Stable Theory of Population. – The Discrete Time Recurrence Model. – The Continuous Time Model. – The Discrete Time Matrix Model. – Comparative Aspects of Stable Population Models. – Extensions of Stable Population Theory. – The Kingdom of Denmark – A Demographic Example. – Appendix. – References. – Subject Index.

Volume 12
R. Gittins

Canonical Analysis

A Review with Applications in Ecology

1985. 16 figures. XVI, 351 pages. ISBN 3-540-13617-7

Contents: Introduction. – **Theory**: Canonical correlations and canonical variates. Extensions and generalizations. Canonical variate analysis. Dual scaling. – **Applications**: General introduction. Experiment 1: an investigation of spatial variation. Experiment 2: soil-species relationships in a limestone grassland community. Soil-vegetation relationships in a lowland tropical rain forest. Dynamic status of a lowland tropical rain forest. The structure of grassland vegetation in Anglesey, North Wales. The nitrogen nutrition of eight grass species. Herbivore-environment relationships in the Rwenzori National Park, Uganda. – **Appraisal and Prospect**: Applications: assessment and conclusions. Research issues and future developments. – **Appendices**: Multivariate regression. Data sets used in worked applications. Species composition of a limestone grassland community. – References. – Species' index. – Author index. – Subject index.

Springer-Verlag
Berlin
Heidelberg
New York
Tokyo

Journal of Mathematical Biology

ISSN 0303-6812   Title No. 285

For mathematicians and biologists working in a wide spectrum of fields, the **Journal of Mathematical Biology** publishes:
- papers in which mathematics is used to better understand biological phenomena
- mathematical papers inspired by biological research, and
- papers which yield new experimental data bearing on mathematical models.

Contributions also discuss related areas of medicine, chemistry, and physics.

Springer-Verlag
Berlin
Heidelberg
New York
Tokyo

Subscription information and sample copy available on request.
Please address inquiries to Springer-Verlag, Heidelberger Platz 3, D-1000 Berlin 33